To my fellow writer Dawn...

DIFFERENT RULES

A whole bunch of marketing rules you're already mastered!

Enjoy...

Chris Northwood
Aug 2023

Advance Praise for **Different Rules**

This work by Chris Wirthwein is practical, instructive, and insightful. I would recommend it to anyone striving to sell their services or products in a B2B market today. It has a unique blend of theoretical reasons why the concepts work with real-life stories and hands-on tactics to implement those concepts. The breakthrough emerges as Mr. Wirthwein brings an old word in the marketing world (differentiation) to new life and applies it to the present-day era of utility. The book is practical enough and detailed enough for marketers with energy and appetite to make a real difference in their marketplace.

—*John Kalthoff*, Marketing Category Leader
Corteva Agriscience

Wirthwein's apt framing of the "commodity conundrum" will delight the enlightened B2B professional. Whether chemicals or cleaning services, leveraging differentiation—one of the most overlooked truths among management teams—is the biggest strategic and financial opportunity in B2B companies today.

—*Maggie Seeliger*, Global Head/SVP Strategy
Energy & Resources, Sodexo

As businesses continue to wrestle with unprecedented change and new competition, marketers need to be front and center in defining their differentiation in the marketplace. While many books talk about the theory of this new reality, *Different Rules* provides a how-to guide with practical tips and real-life applications that marketers can put into action today."

—*Dave Knox*, author of *Predicting The Turn*; formerly
Chief Marketing Officer, Rockfish

DIFFERENT RULES

The B2B Marketer's Guidebook to Product Differentiation

Chris Wirthwein

Rules

noun — a set of advice about the best way to do something: *the rules of product differentiation*

verb — exercising dominating power, authority or influence over: *the product rules the marketplace*

PMP
Paramount Market Publishing, Inc.

Paramount Market Publishing, Inc.
274 North Goodman Street, Suite D-214
Rochester, NY 14607
www.paramountbooks.com
607-275-8100

Publisher: James Madden
Editorial Director: Doris Walsh

Cataloging in Publication Data available
ISBN 13: 978-1-941688-76-2
Printed in the United States

Contents

Preface vii

Introduction: Why a book about differentiating B2B products? xi

1 Why being different is better than being better, or even best! 1

2 Proving your difference: The problem with benefits and the surprising benefit of features. 8

3 Different? Good . . . but not good enough. The importance of relevance 25

4 Differentiating with a category 34

5 Differentiating with the message 49

6 Differentiating with creativity 66

7 15 Differentiation Idea Starters — and 6 to avoid 102

8 Conclusion: It pays to be different! 117

Appendix: General Guidelines about Questions and Questioning 123

References 129

Index 133

About the Author 135

Preface

Narrow subject; brief examination

Not long ago something dawned on me. Scratch that, something hit me and hit me hard. And when it did, it left me dumbfounded, stirred up, and giddy, all at the same time.

Background

For decades, I, along with the agency where I hang my hat had been "doing marketing"—in aerospace and animal medicines, medical devices, industrial abrasives, plant genetics, additive manufacturing, biotechnology, building materials—the list goes on and on—complex B2B products, complex markets, and sophisticated buyers. We'd developed strategies, plans, programs and communication campaigns of all sorts in U.S. markets and around the world. I'd authored books and articles about marketing, been interviewed and quoted by the trade and business press, spoken about marketing at meetings, conferences and universities. And at this point in my career, in fact on this very day, from out of the blue, something struck me I hadn't prepared for.

The result of that whack to the side of the head? *Different Rules.*

Here's what struck me. . .

Perhaps you remember the theme song from a mostly forgotten 1960s movie *Alfie.* It contains this catch line, "What's it all about, Alfie?"

The film didn't do much at the box office. But the theme song took off, became a pop hit and eventually earned its way into the Grammy Hall of Fame. And why wouldn't it? That catch line poses a question for the ages, "What's it all about?" Whether *it* is snow skiing, playing the piano, baking the perfect soufflé, life in general, or B2B marketing, aren't we all just trying to figure out what *it* is all about?

And that *it* is exactly what hit me that day—the answer to the question, "What's it—marketing—all about?" Ready?

Marketing . . . is . . . all . . . about . . .

Differentiation!

Yeah, right, that's it? Marketing is all about differentiation?

Absolutely.

Since you're reading this book, you must be willing to consider agreeing with me. And I hope you become convinced of it by the time you finish reading *Different Rules.*

Thanks for coming along. But first, some ground rules.

Ground rules: How this book works

- Differentiation is a way of being—a mindset way bigger than just marketing. In the very best companies, differentiation embodies the philosophy of the entire organization. That said, because differentiation can produce such powerful business results, you really should get your leadership team, from every area of the company, on board with it. When differentiation belongs just to marketing, the company won't reap the greatest rewards from it.
- To keep things brief, this book deals with a single subject: differentiation. Although mighty in its power, differentiation is not the only marketing task a company must do well. Lots of other factors, not addressed in this book, go into getting and keeping customers, and doing it profitably.

- I intend to stimulate your further thoughts, ideas, and questions. Should those not relate directly to differentiation, you won't see them addressed in *Different Rules.*
- Throughout the book I use the words *product* or *products*. I use them as shorthand for "products and services." In other words, *Different Rules* applies to both B2B products and services.
- Sometimes, I say "we" or "our," as in, "we developed a campaign," or "our concept deals with." The "we" and "our" refer to 5MetaCom, the company I've been with for many years, and my 5MetaCom colleagues.
- From time to time, I reference concepts contained in my previous books, *Brand Busters* and *The People Powered Brand.* Other than a brief appendix, I won't repeat that information. I will direct you to specific chapters or pages where you can find out more.
- Differentiation won't attract every customer or magically make every sale. Some buyers just want lower prices. *Different Rules* stays away from telling you how to cheapen your offer. It heads in the opposite direction: increasing your overall value to customers.
- I strongly believe differentiation is the heart and soul of marketing. And I love what management guru Peter Drucker had to say about the importance of marketing:
  - ✓ "Because its purpose is to create a customer, any business enterprise has two—and only these two—basic functions: marketing and innovation."
  - ✓ "Marketing and innovation produce results; all the rest are costs. Marketing is the distinguishing, unique function of the business."[1]

RULES

noun—a set of advice about the best way to do something:
the rules of product differentiation

verb—exercising dominating power, authority or influence over:
the product rules the marketplace

Introduction
Why a book about differentiating B2B products?

Near the end of the 20th century, Harvard Professor Theodore Levitt gained prominence by inspiring and instructing marketers on differentiation. It all started with publication of his now classic *Harvard Business Review* article: "Marketing Success Through Differentiation—of Anything."

In it, Levitt greeted readers with this startling proclamation:

"There is no such thing as a commodity."

He further declared that the ability to differentiate holds especially true where you might least expect it: with B2B goods and services. Wrote Levitt:

> "Though the usual presumption is that this is more true of consumer goods than of industrial goods and services, the opposite is true." Later adding, "This is true even of those who produce and deal in primary metals, grains, chemicals, plastics and money."[1]

According to Levitt, everything sold in B2B markets could and probably should be differentiated.

His provocative HBR piece unleashed a "big bang." It sounded a call to arms for marketers, especially to those with B2B products to sell. Levitt told those marketing folks to get with it: stop being generic; start *differentiating*. His insistence upon this continued in his later writings, including this comment from his 1991 book *Thinking About Management*:

"Differentiation is one of the most important strategic and tactical activities in which companies must constantly engage. It is not discretionary."[2]

Not discretionary. Did companies heed the call? Where are we today?

Too many choices, too little time.

Search "marketing differentiation" on Amazon books. Good luck reading the 400+ titles that pop up. And as if that weren't enough, here's another book on the subject—this one on how to differentiate B2B products.

Do we really need more instruction on differentiation? Yes.

But why? And why, as I will assert, is the need for differentiation even greater today than in 1980 when Levitt wrote about it?

It's simple. Things have gotten a lot more complex. As marketer and author Tim Williams says: to "get" this, simply look through the lenses of *time* and *choice*.

In the past, people had lots of time on their hands. But what they didn't have were lots of choices. Henry Ford reportedly said in the early 1900s, "You can get any color of automobile you want, as long as you choose black." Funny, perhaps, but true. Ford's purchasing agents surely encountered more of the same when searching for automobile paint, bearings, and other industrial goods to satisfy the engineering specs of the Model T. Williams summed up those days (which continued through the end of WWII) this way: *unlimited time* and *limited choices.*

The great post-war consumer and industrial boom begat a different, more wondrous and complex world—a world bursting forth with innovation, production efficiencies, choice, and plenty of anxiety to go along with it. It continues to this day. And what happened to the quaint concept of unlimited time and limited choice? Gone forever.

These days, you and I, and your buyers, face just the opposite. Your customers wrestle with stress-inducing time pressures multiplied by a perplexing range of product and supplier options—and more information about those options than Ford or Levitt ever would have dreamed. Spoiler alert: Later we'll show you how to get out of a crowded segment and establish yourself as the *only* choice in a category.

Modern day master of persuasion, psychologist Dr. Robert Cialdini, tells us what happens as the time crunch intensifies amid an explosion of choices and information . . .

> ". . . technology can evolve much faster than we can. With the sophisticated mental apparatus we have used to build world eminence as a species, we have created an environment so complex, fast-paced, and information laden that we must increasingly deal with it in the fashion of the animals we long ago transcended."[3]

So, what's your customer or prospect to do? No different than every other human, they:

1. **Take shortcuts.** You do it. I do it, in all phases of life. Running late for work? I don't hesitate to skip breakfast even though I'm fairly sure it's the most important meal of the day. Top marketers understand these shortcuts. Look no further than the concept of "brand." Among other things, brands provide buyers of consumer and B2B goods with a practical shortcut for choosing one product over another.

2. **Accept "good enough."** Voltaire, the Italian philosopher, put it this way in 1770: "Perfect is the enemy of the good." And now, more than ever, people heed this counsel. When deciding what to buy, people think "utility."

   Utility: fitness for a purpose, as in, "will it solve my problem?"

   To be fair, Voltaire didn't use the word "perfect." Instead, he wrote the Italian word for "best." Who pulled the word swap? His "good enough" translator. In Chapter One we'll revisit Voltaire's ideas and examine why, when it comes to marketing, "best" is every bit the trap he warned us against centuries ago.

To avoid *futility*, buyers think *utility.*

Let's look at why and how buyers use utility to make decisions. As backdrop, I'll turn to Tim Williams again for some history. Williams says modern marketing can be categorized into three chronological eras: 1) Interruption, 2) Persuasion, and 3) Utility.

Interruption: To understand this era, travel back with me to the age of the horse and buggy—simpler days of unlimited time and limited choices. How hard was it, for instance, to convince horse and buggy owners of the automobile's advantages? Not hard at all. Marketers needed only to interrupt—get prospects to take a look—and the wonders of the horseless carriage became self-evident. Countless consumer and industrial inventions in this period possessed such superiority. And with little to no competition, they practically sold themselves. The job of the marketer? Interruption—grab the attention of the consumer. But things changed.

Persuasion: Time ushered in increased industrialization, competition and globalization. With it came product proliferation with vastly more choices and options in consumer and industrial goods. Interruption as a tactic lost efficacy. And the marketer's job shifted. Welcome to the era of Persuasion, a period defined by ever more sophisticated marketing—from FABs (features, advantages, benefits), Unique Selling Propositions, branding, comparative advertising, to demographics, psychographics and personas—all of them designed to persuade buyers to choose one brand over another. In these "good old days," marketers held the power. They controlled not only the form and content of information, but also the buyers' access to it. But those days, too, ended.

Utility: Today, thanks to innovation, we all live in the world of limited time and virtually unlimited choices. Welcome to my world and yours—frenetic, overloaded with information, with way too many options, and precious little time. What's a buyer to do? By this point, you already know the answer. They take shortcuts. They think, "my problem . . . me first!" They forget "best." To avoid hassle and save time, especially in B2B, they buy products that do the job and choose them based on fitness for a purpose—utility. Buyers want to know, "Will this help me achieve what I'm trying to accomplish?" And smart marketers recognize the chances of making a sale go way up when they answer, "yes."

Is it any wonder advertising effectiveness researcher John Hallward tells us: ". . . when an ad shows how to use the product, we observe a higher persuasion effect than average."[4] And listen to Anthony Ulwick, author of *What Customers Want.* Writing on market research in *Harvard Business Review* Ulwick said, ". . . customers should be asked only for outcomes—that is, *what they want a product or service to do for them.*" (Emphasis mine.) Interestingly, he wrote those words in 2002.[5]

And Peter Drucker said this, "The most beautifully designed machine is still only so much scrap metal until it has utility for a customer."[6]

And what does all this mean?

Marketers should acknowledge *utility* as the decision-making paradigm ruling our wondrous, distracted, chaotic world today. What's more, it's time to do something about it. What? Become an ally for customers. Move with them into the era of utility-based buying. Help customers understand what's different about your product. Serve buyers, and your sales organization, by explaining how your products fit with what customers want to accomplish. Point out exactly how and why your products and offerings are not at all the same as what your competitors sell.

In other words, *differentiate* by putting utility at the forefront of your marketing.

Sometimes differentiating in this way means you'll talk about features. Other times it means comparing your offer with other products, categories and ideas, which may even include choosing not to buy anything. The rest of this book explores a new marketing curriculum that equips you to master the new marketing era of differentiation and utility. And it's nothing you've learned in school or in any other book.

Final words about the concept of utility

One last time: why do buyers think utility when making decisions? Too little time, too much complexity. To close I'll sum it up this way, in a verse I wrote called the "The Innovator's Impasse."

The Innovator's Impasse

Humanity's flair to innovate
Has surpassed its knack to communicate.
And when buyers can't understand why . . .
The last thing they do is buy.

Because of the "Innovator's Impasse"—the lack of simple, meaningful differentiation—all sorts of technological wizardry have ended up on the junk heap. As marketers we must overcome this. And to that end, there's good news. Communication disasters—failure to explain and differentiate complex products—can be fixed. But before we learn how, let's examine why being *different* holds the key to, you guessed it: differentiation.

People notice and respond to what is different.

On January 9, 2007, in his Macworld San Francisco keynote address, Steve Jobs captured the world's attention and highlighted the *difference* between the first iPhone and every other competing product in just one graphic. On a horizontal axis ranging from "hard to use" to "easy to use" and a vertical axis ranging from "smart" down to "not so smart," Jobs showed all the currently available popular handhelds clustered in the lower "hard" and "not so smart" sectors while the new iPhone sailed alone at the upper reaches of "easy" and "smart." He rounded out his presentation by explaining the utility of the new device.

Too many choices!

Marketing Technology Landscape, The Martech 5000

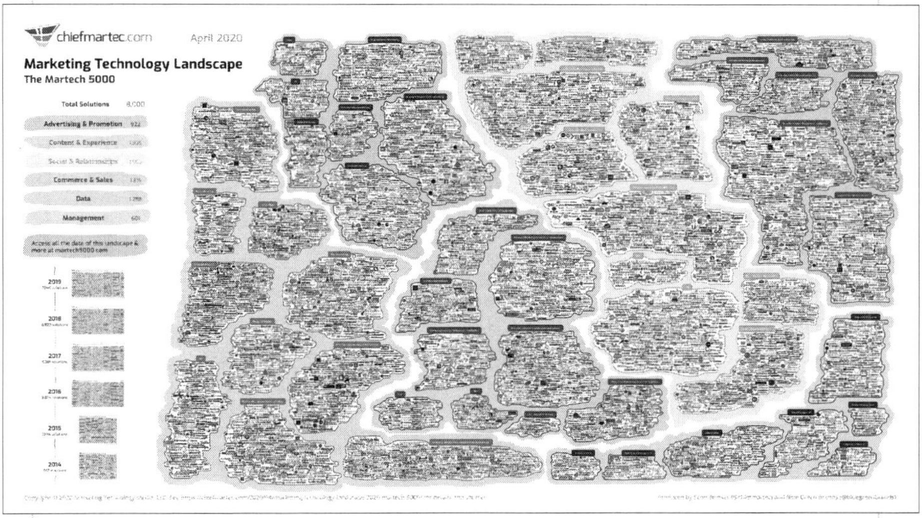

Source: chiefmartec.com

This graphic comes from a story about marketing software. It shows 8,000 or so competitors vying for sales. Which software is the best? Who knows? No one will take the time to evaluate them all. A buyer will take a look at a few and buy one that delivers the utility for the job at hand.

CHAPTER 1

Why being different is better than being better, or even best!

People—*all of us*—have been programmed to automatically, irresistibly and involuntarily stop in our tracks when we encounter something *different*. Credit this hyper-sensitivity to spotting differences for the survival of the human species. This fact hardly requires proof. But just in case, consider this scene:

> It's 2:00 AM in the city. Alone, you walk briskly down a dimly lit, seemingly deserted street. Without warning, out from the blackness of the alley ahead of you, darts a shadow. Instantly your brain signals RED ALERT. It shouts, PAY ATTENTION!! Unconsciously, your body's defense mechanisms leap into action—adrenaline spikes, heart rate increases, blood flow diverts to muscles, breathing shallows, eyesight, hearing and smell sharpen. Automatically, involuntarily, your body has armed itself for fight or flight.

All because you noticed something *different*.

As good fortune would have it, in this made-up example, I'll make the mysterious apparition a harmless alley cat. Disaster averted, but point made, hopefully. And the point is this: to animals—*all* animals, including humans—*different* stands out. *Different* grabs attention. *Different* unconsciously engages our brains and bodies; we can't prevent or stop this. *Different* engages our mind and body at its deepest animal core. When we encounter something *different*, we pay acute attention. You noticed the different *italic* typeface I used, yes?

So, what happens when a prospective customer becomes exposed to something different about a product you market? Do they react like the person in our late-night example above? Not precisely, but for our purposes, the answer is still yes. While a product difference may not trigger an adrenaline rush to the body, researchers tell us *different* will be noticed. It will seize the prospect's attention.

In his groundbreaking book, *Gimme! The Human Nature of Successful Marketing*, market researcher John Hallward says it this way: ". . . humans are wired to detect irregular things."[1]

Neuroscientist and author Dr. John Medina said something similar in his best-seller *Brain Rules*: "We don't pay attention to boring things."[2]

Harvard Business School's Dr. Michael Porter, creator of the Five Forces of Competition model urged marketers to give up on competing to be the best and instead "compete to be unique."[3]

And how does being unique benefit the marketer? Harvard Business School's Youngme Moon, author of *Different: Escaping the Competitive Herd*, said this: ". . . the way to think about differentiation is not as an offspring of competition, but as an escape from competition altogether."[4]

When it comes to human behavior, it should come as no surprise to you that our animal defense mechanisms do not trigger when we're exposed to something that's better or even best. Only *different* gives us a free ride deep into the psyche of our customer—in ways that being better or even best can never accomplish.

The book *How People Learn* reports on two studies that reveal the deeply ingrained human impulse to recognize differences. In the first study, 6- to 8-month-old infants were shown photographic slides of common household items. Each slide contained either two or three items (a comb, foods, scissors, cat, etc.) Half the infants saw a series of two-item slides. The other half saw three-item slides. As more and more slides were shown, the infants became bored. Their "looking times" dropped by 50 percent. Next, came a series of slides that alternated between two and three items. "Looking times" went back up. Simply showing a different number of items seized the infants' attention.[5]

A separate study of 5-month-olds measured "surprise reaction" when an item was either added or subtracted. Infants were shown two non-identical, random objects. Researchers dropped a curtain and either added or subtracted an object, then pulled up the screen. In both cases—more items and less—infants looked longer when the newly revealed setting contained a *different* number of items.

Returning to the world of marketing, Harvard's Joan Magretta makes a strong case for utility in her article "Stop Competing to Be the Best:"

> ". . . think about all of the industries in the economy. In how many does the idea of "being the best" make real sense? The best hotel for one customer is not the best for another. The best sales encounter for one customer is not the best for another. There is no best car. There is no best art museum."[6]

Finally, consider this. Generally speaking, by proving you're better or best who wins? When you convince them of your superiority, you make a sale. You win. You get money. And yes, I've heard all the talk of win-win. But when you make a sale, one person receives a guaranteed, immediate win: the seller. The other win—the one for the customer—comes later. Maybe. Can you see why customers look skeptically at claims of better or best?

On the other hand, who wins when you show the utility of your product? Customers. They get something or achieve something they care about. This makes you relevant to them—which is something you should dearly want to be. (More on relevance in the next chapter.) Not to get preachy, but I view showing utility as the moral high ground. It says you put the customer first. And it says so way better than sloganizing with something as dubious and unimaginative as, "We put the customer first."

But what if you really are better or best? Go easy on using the words best or better—unless you're comparing to another of your products and you're helping a customer understand your product's utility. Example: "Our product A performed better in high temperature applications than our product B." In addition to helping the customer, admitting a weakness almost always builds credibility.

As a rule, use facts related to features, rather than your opinions, to assert advantages. Here's how: "World's lightest carbon-based filament for under-sea environments"—a true, provable, non-opinion statement. Stating the feature this way makes the filament different from others. Only one filament, in a category such as undersea, can be the lightest. Technical people who hear a fact-based statement like this will recognize the filament's lightness as a *difference*. Furthermore, they will intuitively conclude your filament has utility for weight-constrained under-sea operations. This is their opinion—not yours. And thus, they will be more likely to believe it.

Another type of opinion that's okay to share is one from a recognized authority. And who that authority happens to be will depend on the situation. Who are the authorities in your field? These are the folks whose opinions are okay to use—it could be a person, a university, a third-party company, a testing organization, an NGO, etc.

Finally, data backs up the business value of being seen as different and unique. Researcher John Hallward reported that as perceptions of uniqueness for products went up, so did purchase loyalty.[7] The more unique and different a brand rated in his tests, the greater the tendency of buyers to re-purchase. One interesting caution, however, arose in his studies. Increased loyalty did not extend to brands ranked at the absolute highest level of uniqueness. Those brands came to be seen as boutique or specialty goods—products only to be bought infrequently. (Think of expensive champagne or a tuxedo or ball gown for an ultra-special occasion.) While I know of no such data supporting this finding for B2B goods, it seems logical this caution should also apply here.

The bigger point remains: being different provides a powerful marketing edge because the brain instinctively spots differences. In doing so, it signals us to become alert and pay attention. How important is the concept of "different" to how the human mind works? Consider this: the index to Dr. Medina's *Brain Rules* lists no fewer than 78 entries containing the root word of "different," and a total of zero for "better" or "best."[8]

We don't have to work to cause someone to notice and pay attention to something different. The human mind and body do this automatically, likely from birth. They do the work for us. Talk about effectiveness and cost efficiency! When we highlight what's different, we spontaneously cause the customer's mind to do our marketing work for us.

In *Gimme!*, Hallward concurs with this:

> "Detecting irregularities has become well established in our genes and will not quickly disappear."

And this:

> ". . . if we want to get attention and be recognized, we have to be irregular (stand out from the crowd)."[9]

Human nature, animal instinct, involuntary response—that's why being different is better than being better or even best. And that's why being different holds the key to differentiation. Seems self-evident, right? Yes! So, we're done?

Not exactly. Remember how medieval scientists believed the earth sat at the center of the universe only to be proved wrong? Get ready to unlearn one of the "great" lessons of marketing as Chapter Two takes an in-depth look at features and benefits.

You've nearly finished reading a chapter unlike any other . . .

. . . which makes it different from all others. And here's the moral of Chapter One's story: Be different and you'll end up better off than if you were better, or even best.

And who knows, if you're not already, maybe you'll end up being #1, which just might be the best different you can be!

A few tips about being different . . .

- "New!"—the old standby

  Perhaps nothing better conveys *difference* than "new." People intuitively know new means different. And they display a huge appetite for wanting to learn more about what's new and different. Anytime you have something new—a new product, new information about an old product, a new promotion, new anything—say so. New makes you different from everything else that came before. Yes, yes, I understand new can also mean untested, unproven or untried. To overcome that hurdle, most B2B products brought to market arrive with factual support—test data, certifications, regulatory approvals, etc. Be ready with that information. But most of all do this: When you have something new, point it out.

- Internal forces *over*value better or best and *under*value different.

In my experience, sales, marketing, manufacturing, R&D and other internal groups discount the inherent marketing power of being different. I understand. Their worlds are overwrought with pursuit of better or best. That pursuit drives business success and that's commendable. But, when marketers make such superiority claims, skeptical buyers tune out.

An important part of a marketer's role is to educate internal audiences. The primacy of differentiation to marketing success should be part of your teaching curriculum. Use the concepts in this book to educate. Better yet, build a corporate ethos that works to understand and explain differences you provide to customers with your products and services. Companies that embrace this belief display an incredible resolve for discovering and defining the true utility of their products. This embrace of what your customers want will make you different from those who boast about superiority. Best of all, the marketplace will reward your company for this valiant spirit.

- Use the "What's different about . . . ?" exercise.

Here's a super simple group exercise I use in marketing workshops with clients. Give everyone a pad of 3"x 5" sticky notes. Set a timer for three minutes. Ask this question of the group: "What's different about __________?" Fill in the blank with whatever dimension you're seeking to differentiate on. For instance, you could fill in the blank with such things as:

... the way we design our products vs. company X?

... the way our product works vs. others?

... the new way we manufacture (or test, package, support, distribute, etc.) our product vs. others (or vs. the old way we used to do it, etc.)?

... what our product does, or can do, vs. others?

... or just about anything else that comes to mind.

Tell everyone the point is to get ideas flowing. Explain you're after quantity; even not-so-good ideas are A-OK. Start the timer and ask people to write one idea per sticky note. When the timer goes off, go around the room. Have folks read off their ideas and stick them onto a white board or flip chart. From there you can group the ideas into broad general themes. Your people can then hash through the idea groups, critiquing, refining, expanding, and editing as they go, theme by theme by group. As an alternative, you could do this exercise with a panel or advisory board of customers. There's really no right or wrong way to do this exercise. In my experience, it's a good ice breaker that gets people thinking about differentiation.

▪ Being different just to be different is just a bad idea!

Standing out by being absurd takes no strategic thought or creativity. And it risks making you, your brand, your company or product look foolish or at best, unserious. Creativity with relevance differentiates you in positive ways. Chapter Three talks in detail about the imperative of relevance.

CHAPTER 2

Proving your difference

The problem with benefits and the surprising benefit of features.

In Chapter One, I used a filament example. You'll recall the material possessed a key difference. It weighed less. This feature can be proven by weighing it. It means the claim—lightestness—is *de facto* (in reality, actually) true. Fact, not opinion.

Features such as weight or speed (and many others) in B2B marketing can be huge differentiators and undeniable proof points for your product's utility. And this ability to substantiate claims and address utility is the surprising benefit of features.

So, let's delve deeper into psychology and explore new ideas for how to boost marketing effectiveness by challenging some age-old assumptions about features and benefits. To do this, we'll discuss an all-important principle: putting reasons—proof—behind your claims of being different.

The essence and application of the "proof principle" comes down to two words: ***features*** and ***because***.

Marketers should know to provide a "because" to support claims. And in fact, a because is essential to substantiating a difference. Shortly, I'll share astonishing behavioral research that sheds light on this and shows that any reason, even a weak or irrelevant one, is better and more persuasive than no reason at all.

But first, let's refine our understanding of features and benefits, starting with definitions. You'll find the **feature** definition to be pretty standard stuff. The **benefit** definition, however, includes a new way—*two* new ways, in fact—of thinking about them.

Features and benefits: definitions

To get this started, I'll use a simple example. Back in horse and buggy days, a unique feature of the first automobiles was speed. And that one feature meant a traveler could get "there" in less time—an extremely specific type of benefit.

But today, in the U.S., where speed limits restrict us to 70 miles per hour or less, no passenger automobile can claim a unique benefit of getting you there in less time. Speed has become a generic claim in the passenger automobile category. Accordingly, a claim of faster provides no differentiation. In certain types of cars, however—drag racers, for instance—speed is a differentiator. But recognize, it is not a benefit. Speed is a feature. How so? Time for definitions:

> **Feature:** *A tangible, observable, measurable characteristic.* From the Latin "tangere"—*to touch,* or, *able to be touched or felt.* A feature describes an object's nature, what it *is*. Examples include attributes such as weight, length, color, how it's made, specific gravity, tensile strength, what it's made of, price, density, how it works, and myriad others.
>
> **Benefit:** *A positive result.* Benefits express what can be done or accomplished with an object—for our purposes, a product or service. Benefits describe outcomes, what an object *does.*

Now the new part—a new way to think about and define benefits. *Two* types of benefits exist: 1) ***direct* benefit** and 2) ***derived* benefits.**

> **Direct benefit.** For example, a diamond-coated steel cutting bit lasts longer than an uncoated one. Longer life—durability—is a positive outcome resulting directly from the diamond-coated feature, hence the term *direct* benefit.
>
> **Derived benefit.** Going further, durability helps to deliver other, extrapolated, or *derived* benefits. For instance, since a more durable bit cuts a greater quantity of material before needing to be replaced, it can save its owner money on replacement parts and

labor cost to replace them, resulting in a lower total cost of ownership. A more durable tool that requires fewer work stoppages to replace worn components can also increase uptime and manufacturing throughput. Finally, less total cost, plus increased uptime and throughput can ultimately deliver greater profit to the business. And all of these ROI-related benefits can be further extrapolated into an array of psychologically-related, derived benefits—softer values such as peace of mind, pride, job satisfaction and many more. *Derived* benefits stem from *direct* benefits.

Derived benefits: higher ground? Or marketing abyss?

The lengths to which we can take derived benefits border on silly. Can our product make you a better manager, get you a raise or promotion, help you achieve self-actualization? As marketers where do we stop?

In user-interviews we've conducted over decades for all sorts of technical and scientific products in all sorts of B2B markets, buyers consistently decry the absurdity of *derived* benefits. I've heard customers use words like this to describe marketing claims of derived ROI or emotional benefits: "exaggerated . . . incongruent . . . silly . . . out of touch with my world," and even "arrogant." I'll never forget the customer of a client who once told me something like, "Who the hell are you to tell me I'm going to feel all warm and fuzzy inside when I use your product? Get real. Do you realize how hard this damn job is?"

On the other hand, customers have told me, without prompting, that using a certain product does indeed give them "peace of mind." Wait a minute; how can this be? How can the very words that cause annoyance when a marketer uses them be the exact words a customer uses to describe a personal experience? First off, understand there is no requirement for customers to be logical. Get rid of that notion. Customers get to "be" however they choose. So, what gives?

It's okay for customers to share their emotions. It's okay for them to tell you how they feel or how much economic return they get from a certain product. But when you as a marketer tell them what their business results or emotions should be, you just stepped over the line. Derived economic or emotional benefits customers see or feel are their call. Not yours.

Here's my advice. When it comes to economic- or emotion-based, derived benefits, don't say it yourself. Instead, have someone else say it —a well-respected third party, a named customer or at the very least an unnamed customer. Something like this: One plant manager of a 24/7/365 operation recently told us, "I'm saving 30 minutes per shift changeover." Again, let *them* say it, not *you*.

And here's a real-world example. We once created a marketing campaign for a professional service staffing firm. They provided "on call" clinical workers to hospitals for temporary assignments. Demand for their service always outstripped supply. The bottleneck? Finding people. Consequently, they pumped the majority of their marketing dollars and effort into recruiting workers. Their target: newly credentialed professionals with 0–3 years of work experience.

To attract these employees, we created a campaign that used a derived emotional benefit, expressed by someone else and not the company, as its appeal. Ads featured a photo of a young professional on the phone. The headline (in quotes) said, "Mom, I'm really glad I took the job . . ." The copy let us "listen in" on her conversation with Mom. Now what daughter wouldn't want to call Mom to share good news on the job front? That joy, satisfaction, and emotion is what the campaign captured and communicated.

I wish I could take credit for creating the idea. In truth, I uncovered it while doing background research interviews with my client's satisfied employees. Several reported they couldn't wait to call home and tell Mom how happy they were about taking the job. That insight became the headline of an effective campaign. What's more, it taught me the lesson I just shared. When it comes to derived benefits, have *them* say it, not *you*.

Benefits invite debate . . . features not so much

While your customer may debate or doubt benefits you claim, the same cannot be said for features. Features can be tested and proven in ways acceptable to customers. Features can be shown to be absolutely, empirically true. Benefits often come down to an argument of opinions.

I once worked on a new formulated liquid product that contained two active ingredients. Other products in the category contained just one. This unique feature—two active ingredients—is an undeniable truth. And in this case, it delivered a significant differentiator for our client's product. In fact, in qualitative interviews, prospective customers told us how much they liked this feature. Many said they reasoned the new product would work better *because* it contained two active ingredients, instead of just one. Had we instead shouted, "OUR PRODUCT WORKS BETTER" in our communications, we might have invited debate from users of competitive products. But we didn't. We started with mention of the feature, indeed a fact: this product contains two active ingredients. Customers could see this for themselves on the package label. Rather than provoking a debate, salespeople told us that opening with the feature had a way of engaging customers in an easy, non-confrontational conversation about what was different about the new product.

Benefits or features: Which are more persuasive?

Marketing experiments with consumer goods show both can be persuasive, depending upon the situation. For instance, is the purchase to take place in the near or distance future? And is the mindset of the buyer oriented to theoretical, abstract thinking or tangible absolutes, so called "concrete thinking?" Several experiments shed light on this.

A study published in 2015 reported on four experiments conducted with several different considered purchase (not impulse purchase) products: a notebook computer, a tablet computer, a GPS device, and a smartphone. The tests compared persuasiveness of benefit-oriented appeals versus feature-oriented appeals for a product that was to be purchased in the near future (one week) versus the distance future (six months). The study also examined persuasiveness of appeals when participants were induced to think one of two ways: abstractly or concretely. Results showed:

- When a purchase was planned for the near future, feature-based and benefit-based appeals were equally persuasive.
- When a purchase was planned for the distant future, benefit-based appeals were more persuasive.
- When concrete thinking was induced, feature-based appeals were found to be more persuasive.
- When abstract thinking was induced, benefit-based appeals were more persuasive.[1]

Does the mindset of your audience lean toward rules-based, tangible thinking? Or do you appeal to more visionary, theoretical thinkers? Ask yourself, what is the tendency of our audience?

These experiments suggest that the more a purchase is thought of as real vs. theoretical, the more persuasive tangible features become. This research implies that theoretical, visionary appeals can be persuasive early in a customer's decision making process. But in the later buying stages, when the actual purchase nears, the more persuasive the product's tangible, feature-based appeals become.

Use of derived benefits creates risk of becoming GENERIC

Most marketers "get" features. But plenty of B2B marketers get tripped up on benefits. How do I know this? Because I, along with my clients, got tripped up by them for years. We got tripped up by using *derived* benefits when we should have been using *direct* benefits, and the corresponding *features* that support them.

What's bad about using derived benefits?

To start with, derived benefits don't address customers' deep-seated desire and expectation for *utility*. Derived benefits don't speak to hard-edged practicality—the utility B2B buyers expect in considered purchase products for their business. Yes, derived benefits speak to loftier ambitions: profit, ROI, peace of mind, wealth, early retirement, etc. But they don't address wants and desires of customers who simply yearn for a product fit to the purpose of accomplishing a well-defined task. In many cases, utility will be the deciding factor in B2B buying decisions.

There's another problem with derived benefits—one that's worse. Derived benefits almost never differentiate. Just about everyone uses them. And when you use them, you become like everybody else: undifferentiated. And what's another word for undifferentiated? GENERIC! Does the thought of your intentional marketing efforts making you generic send shivers down your spine? It should.

The "Big 3 in B2B"

I sometimes joke about what I call the "Big 3 in B2B." These are three worn-out *derived*, extrapolated benefits that can be ascribed to just about any and every B2B product or service. You know them. Unlike *direct*, product-specific benefits, you see these generic claims made every day. This should be a huge clue to you that they are not and cannot be differentiators.

I'll ask you to stop reading for a moment to see if you can guess the "Big 3."

Time's up . . . here they are, the "Big 3 in B2B":

1. Save time/hassle
2. Save money
3. Make money (ROI)

When you use any one of the "Big 3"—and many folks use them all at the same time—you do not differentiate. You have been warned.

Am I saying customers don't care about saving money, making money, saving time and hassle, or experiencing peace of mind, job satisfaction and pride? Not at all. They absolutely care. But it ordinarily requires a lot more than simply purchasing a product from a supplier to make business gains of this sort. And buyers know this. Marketers who make these types of claims can be seen as cliché, out of touch, or even arrogant. Bad. Very bad! And don't miss the marketing point. You will have a hard time differentiating on derived benefits.

So, marketers should never use derived benefits, right?

Wrong. There's a proper time and place to talk about economic or emotional derived benefits. The proper time and place are ***after*** your customer understands you are *different* and ***after*** you satisfy his or her desire for *utility*. Furthermore, it's usually best to deliver these messages one-on-one, for example, your sales and technical people sitting down with their key people in a consultative meeting of the minds centered on helping the customer make a good business decision. That's the time to run the numbers on saving time, saving money, and making money. But it almost never makes sense to start there.

Final words on benefits and how to use them to differentiate:

Stop (over)using *derived* benefits. They make you generic.

Start using more *direct* benefits, supported by your unique features. Together, they make you different.

PROVE IT!
Case Study #1: The copy machine

As a marketer of a technical/scientific product, you no doubt know your audiences. My guess is they possess both sophistication and technical savvy. They may be highly experienced and street smart, highly educated, or both. And when it comes to buying things, they almost certainly don't like to be "sold." They usually look skeptically on marketing claims and won't just take your word for it. They expect and frequently demand proof.

So, what's a marketer to do? Easy: Prove it! But how? How much and what type of proof—the "because" support for your marketing claims—do you need?

Before you answer, remember the most important thing to prove is not that you are better, or best. The most important thing to prove is that you are *different*.

And this is where features, again, prove their value. Let's take the case of our lightweight filament. The "because" for its lightness comes in the form of features—things like material composition or manufacturing methods. But are features, as proof, good enough? In other words, how good must our reasons, our proof, be? Let's find out.

Perhaps you've heard of the famous copy machine study from Psychology 101. It showed just how good a reason must be in order to persuade people to do something. More than that, it changed our understanding of human behavior. It showed we don't need amazing, irrefutable reasons in order to persuade. Believe it or not, a reason of any kind—even an immaterial *or irrelevant* one—can be good enough.[2]

Don't get me wrong. I don't suggest you go in that direction. But I do want you to learn a lesson about human psychology. When you do, I suspect you'll conclude you have plenty of reasons "good enough" to prove you are different.

Here's how the copy machine study worked:

> A Harvard professor asked her research assistants to cut in front of people waiting to make photocopies at the library. When they spotted a line at the copy machine, a researcher would approach an innocent bystander and ask one of three versions of very specifically worded questions:
>
> **Version 1:** *"Excuse me, I have five pages. May I use the Xerox machine?"*
> A request only; no reason
>
> **Version 2:** *"Excuse me, I have five pages. May I use the Xerox machine, because I'm in a rush."*
> A request with a real reason
>
> **Version 3:** *"Excuse me, I have five pages. May I use the Xerox machine, because I have to make copies."*
> A request with an immaterial reason

The results:

Version 1: 60% of people let the researcher skip the line (no reason)

Version 2: 94% of people let the researcher skip the line (real reason)

Version 3: 93% of people let the researcher skip the line (irrelevant reason)

What can we learn from this? Having a reason improves persuasion. Notice versions 2 and 3 both contain the word *because*. This simple word—*because*—seems to make all the difference.

PROVE IT!
Case Study #2 (more on how good we must be)

Three business professors sought to discover the marketing power of *because*. Rather than study the influence of reasons on compliance with a simple request (making copies), they examined the effect of providing additional attributes on product preference. Writing in the pages the *Journal of Marketing Research,* authors Carpenter, Glazer and Nakamoto detailed their findings.

In one of their experiments, participants were given descriptions of eight hypothetical brands of down jacket, one of which was differentiated by an irrelevant attribute. No price information was provided. Seven of the brands were said to contain "regular down filling." The eighth brand contained "Alpine class down fill." No explanation was provided about the meaning of any attribute. Subjects were asked to rate the brands from least preferred (1) to most preferred (10). Average rating for a product with "regular down filling" was 3.1. The "Alpine class down fill" product rated 9.1, a statistically significant difference.

In a separate test cell, participants were provided with a paragraph of explanation about the attributes and their functions. "Alpine class" was described as goose down. "Regular fill" was described as goose down or a mix of goose and duck down. Subjects were told the type of bird the down comes from "does not make a difference" in performance. A later "manipulation check" revealed that subjects acknowledged Alpine class to be irrelevant. Product rating for a "regular fill" product averaged 4.3. The "Alpine fill" product averaged an 8.4 rating, also a statistically significant difference. The authors concluded, "Even if subjects acknowledged that the differentiating attribute is irrelevant, they rate the differentiated brand more positively than the same brand without it."

The next part of their study tested pasta, audio equipment, and down jackets. It also disclosed prices to the participants. The differentiating attribute included for the down jacket was the same as above. Other differentiators included: "regular signal processing" vs. "studio-designed signal processing" for audio equipment and "regular style" vs. "authentic Milanese style" for pasta. Here's what the experiments revealed:

- **When irrelevance of the differentiating attribute was <u>not</u> revealed:** adding the attribute increased valuation of a brand priced higher than average and even more at a *premium* (highest) price. Adding a unique attribute did not boost perceived value of a brand priced lower than the average.

- **When irrelevance of the differentiating attribute was revealed:** adding the attribute increased valuation of a brand at a higher price but not for a lower or premium-price brand.

Being different pays.

What does this series of experiments tell us? Being different pays. "The prevailing view is that successful product differentiation requires distinguishing a product or brand from competitors on an attribute that is meaningful . . ." Yet, the authors' experiments yielded a radical conclusion, ". . . differentiation on an irrelevant attribute—'meaningless' differentiation—can create a value difference between brands and, in the process, a meaningfully differentiated brand." They surmised, given the mere presence of an attribute, "buyers infer it is valuable—even when told otherwise."

The paper stated, "The differentiated brand will be more favorably evaluated than an undifferentiated one." And finally, ". . . uniqueness can simplify decision making, convey novel information that receives greater weight, or affect the comparison process in such a way as to disadvantage brands lacking this attribute."[3]

There you have it: two completely different experiments, eerily similar results.

Do these findings apply to B2B goods? It's anybody's guess, but in the absence of contradictory evidence my gut tells me the findings would still hold true. Human nature is human nature.

I'm not telling you to simply pluck irrelevant reasons out of thin air and use them to support your marketing. Your audience is likely too sophisticated and skeptical to go for that. B2B marketing demands more than persuading people at a random copy machine or inducing the purchase of a handful of consumer goods.

The moral of the story, the lesson for marketers is this: You must have a *because*; You must provide a reason for your claimed difference and benefits (such as a unique feature); You must provide a reason why customers should believe, for each key premise or message point you make.

PROVE IT!
Case Study #3: Welcome to your world!

So much for psychology and market research experiments. Here's a real-world example on using a feature to prove and differentiate.

Some years ago, we worked on a project to develop a differentiator for a highly technical, decades-old product. It would be fair to call the product "sleepy" and overlooked. Sales and share were okay, but flat for years. Along came a new product manager. Her mission: revitalize and grow the product. In a word: differentiate! Our audience target: highly educated decision makers, some with doctorate degrees. Our assignment: come up with a differentiator to revitalize the brand.

First step: We conducted market research—about a dozen qualitative discovery interviews with the audience. It gave us highly informed, voice of customer insights. For more on methods for how to do this, refer to

"A few tips about learning what's relevant to your audience" at the end of Chapter Three.

Next: Informed by insights, the 5MetaCom team developed a list of several dozen candidate differentiators, all scientifically valid. Some consisted of derived economic benefits. Others communicated direct benefits that addressed specific applications or problems. Still others consisted solely of features. (Note: we developed our definitions of *direct* and *derived* benefits after this example occurred.)

From there, we set up a market research study to identify the most powerful differentiating attribute—defined as a characteristic considered both unique and highly valued.

We first asked participants to rank order the group of feature and benefit statements—high to low—from one to 25 on "importance to you." This would be similar to asking you to rank order on importance various attributes about a car purchase: stylish looks, fits my budget, good gas mileage, electric powered, great handling and performance, 100,000-mile warranty, etc.

After respondents ranked statements from most to least important, we asked them to rate each statement on a 1–10 "uniqueness scale" (with 1 = not at all unique: other companies or products say this, and 10 = highly unique: I can't think of any company or product saying this). Then we calculated a "differentiation score" for each attribute, using an equation that cross-factored importance and uniqueness.

As you might expect, top-ranked statements for importance scored low on uniqueness. These included things such as value for money, good performance, ROI, etc. So, they had not considered them unique or differentiating. Lurking near the top of the importance rankings, however, stood a somewhat neutral, perhaps even bland, feature: "I know how it works."

Interestingly, this statement hit all the right notes, scoring a near perfect 10 on uniqueness—the highest ranking of all 25 or so that we tested. Truthfully, I had almost forgotten we had added it to the list. Like most marketers, I had been trained to rely on benefits to make our case. Earlier I told you I had been tripped up by benefits. Here's when I began to learn the power of features to differentiate!

The audience did not rank "how it works" #1 in importance. But they ranked it very near the top—another "ah-hah" for me. We had found our "different."

A scan of competitor messages proved out the findings. No competitor talked about the science behind how their product worked. Both the #1 and #2 brands in the category focused their message on profit and return on investment. Crazy, but what each said about its product was nearly identical to the other. In essence, each claimed "we're best." I sensed the market was ripe for a disruptor—a product with something different—a unique story about how it worked.

We developed a campaign that told the mode of action story. In doing so, we established our difference. It resonated with the audience. The product rose from #3 share to #1, which gave us an additional differentiator—the most used and preferred product in the category.

Perhaps you won't be surprised to learn what we did next. After becoming #1 (a quantifiable feature-based differentiator), we created and launched a "tribute" campaign. It said "thank you" to the thousands of professionals in the industry who had made our client's product the industry leader.

Then, in our follow-up campaign, we let the voice of the customer speak to that difference: category leadership. Our audience (and likely yours) knows their peers aren't stupid. They know when more of their peers choose a particular brand more than any other, it must deliver utility. I realize people tend to be skeptical. They recognize marketing as what it is—an attempt to sell. Still, the wisdom of the crowd speaks loudly and credibly to animals, including those with a human psyche.

Everybody knows only one product can be #1. And that makes a #1 product supremely different.

What started as a search for how to be different grew into a highly successful, multi-year campaign that boosted a decades-old product to the top and kept it there for years.

Are we telling you to rush right out and put together a "how our product works" campaign? No. But you should, with all haste, develop or discover your differentiator. This difference could be a direct benefit. If so, back it up with a reason—a "because"—which in many cases will be a feature. On the other hand, your difference may be a feature. If so, go with it. Chapter Seven gives you lots of ideas on where and how to go looking for your difference. But before then, let's summarize:

> To differentiate, you must be different. You must also prove your difference with a *because.*

But to differentiate well, there's one more thing you must do. You must be relevant, the topic of our next chapter.

A few tips about using features and benefits . . .

- In mature categories, features can make the difference.

  The initial entrant in a category can often capture the market with direct and even derived benefits. Imagine you're launching the first automobile. You could credibly make a differentiating claim such as "get there faster." However, once the category became crowded, and with the advent of speed limits, speed became generic; every product could make the same claim. Derived benefits often become generic when a category gets crowded. So, what's a marketer to do? Emphasize features. We see it all the time with pharmaceutical products: "The only once-***daily*** treatment for . . ." Later, we shouldn't be surprised to see some other competitor launching with yet another differentiating feature, "The only once-***weekly*** treatment for . . ." In highly mature categories, features are often among the very few differentiators available to a marketer. Use them to your advantage, especially in highly mature categories.

- **Technically oriented buyers find comfort in features**

  A paper cited earlier mentioned how people with a rational, concrete mindset are more persuaded by features rather than benefits. My experience bears this out. Technical buyers tend to be skeptical about benefits and find reassurance in concrete, verifiable features.

- **Features can speak to direct operational benefits**

  Imagine you market a dry ingredient to food manufacturers. Your key difference: concentration. The product has a high weight to volume ratio. A ton of it takes up half the space of other products in the category. Imagine what this feature means to a plant manager: less volume of inventory to keep on hand, fewer storage bins required, fewer shipments per month to take in at the loading dock, etc.

  These direct benefits, coming from a single feature, can tell a compelling story around operational issues. Start with the feature and then explain how it delivers direct operational benefits.

- **Which comes first: feature or benefit?**

  Earlier, I used the example of a product with two active ingredients. In market research prior to launch, we evaluated several concepts: some led with the feature (two active ingredients), others began with a benefit (product performance). Feature-first turned out to be the clear winner. We used the two-ingredient difference —a feature—as our opening line in launch communications and in face to face selling. It worked. This doesn't mean you should always lead with a feature. When in doubt, test.

- **Benefits resulting from a specific feature depend on the audience**

  In the concentrated ingredient case above, I talked about operational benefits of this unique and relevant feature. High concentration may also deliver different benefits to others in the organization. For the sustainability director, the feature could equate to a greener supply chain: fewer deliveries, less trucked weight into the plant and less fuel used in transportation. Finance could realize savings in capital equipment: less investment and floor space tied up in ingredient storage capacity. You get the idea. Different audiences within a customer organization may appreciate your unique feature for many different reasons. Be ready to explain distinct benefits resulting from your difference to multiple audiences.

- Credibility of benefits comes down to "says who?"
  Features such as size, weight, strength and color can be proven. The same cannot be said for benefits, especially derived benefits of an economic nature: profit, cost savings, return on investment. Derived benefits like these rely on many assumptions: energy cost, labor rate, cost of funds, etc. Businesspeople, as they should, look skeptically on assumptions offered up by a seller. When you can, find others outside your organization to vouch for your case: a well-known customer, an industry consultant, a regulatory authority, a university professor, etc.

CHAPTER 3

Different? Good . . . but not good enough.

The importance of relevance.

Ever read Shakespeare? In *Twelfth Night,* a mischievous maid, friend to lady Olivia, plays a prank on Malvolio. He's a buffoonish oaf who secretly pines for the much sought-after Olivia.

The maid pens and delivers a phony love letter to Malvolio, signing Olivia's name. In the note, "Olivia" professes a secret love for Malvolio. She asks him to prove his love for her by appearing at a formal gathering dressed in yellow stockings and crossed garters. And so, he does, looking clown-like and standing out conspicuously.

As if that weren't enough, Malvolio loudly addresses Olivia. He shouts passages from the bogus love letter, which Olivia, of course, knows nothing about. For all the world to see, and quite proud of himself, Malvolio stands out. Compared to Olivia's competing suitors, he is the picture of "differentiation." To Olivia and the court, the boisterous Malvolio appears a madman. He is captured, taken away and confined as a lunatic. So much for differentiation.

But here's the moral of the story for marketers: When pursuing customers, simply standing out—being different—isn't enough. To win, you must also be *relevant.*

Hold on. In the previous chapter, we learned irrelevant reasons can persuade. So, now I'm telling you to be relevant? Absolutely. Think about it. The copy machine experiment did not involve selling a product. The other study used consumer goods, items completely relevant to participants.

Besides, nobody, especially me, ever said you should knowingly be irrelevant in your marketing.

Market researcher John Hallward studied the effect of relevance on purchase. He asked users and non-users of various products to rank them on five factors: familiarity, uniqueness, relevance, popularity and quality. Participants ranked goods they used higher on all five factors compared to products they did not use. Hallward concluded, "relevance is the biggest difference between users and non-users."[1]

But what about your situation as a B2B marketer, dealing with time-crunched buyers and complex products? How important is relevance here? Very important, because relevance tells the brain to pay attention. According to neuroscientist Dr. John Medina, "the brain selects meaning-laden information for further processing and leaves the rest alone."[2]

Meaning-laden information—relevance—what a powerful idea. But please understand who gets to decide what's relevant. It's not your scientists, your technical folks, your R&D people, not me and not you. It's the customer.

Relevance: A simple idea. Not so simple to pull off.

Being relevant draws people in. Being relevant says to customers and prospects that you understand their world. It implies you care. In addition, being more relevant than your competition can create differentiation. Relevance creates a win-win for buyer and seller.

Let's dig deeper. Using you as an example, *relevant* is anything important to you. Relevant is what you value, care about, and have interest in. There's nothing terribly complex about what's relevant and what isn't. But when it comes to marketing B2B products, being and staying relevant can get messy, especially when you consider internal and external issues, complexities and distractions.

Remember this:

Relevance = appealing to self-interest

And self-interest drives us all—me, you, every other marketer and your customer. Great marketers get this. They display a knack for overcoming pre-conceived notions and setting aside their own self-interests. They learn to understand and address self-interests of others. They don't let their own views or those of internal peers and superiors get in the way of communicating with customers on their terms. Sure, good marketers pay attention to internal influencers. But they execute marketing—positioning, messaging, branding, and promotion, in ways relevant to the customer.

Dale Carnegie, author of all-time bestseller *How to Win Friends and Influence People* had this to say about relevance:

> "The royal road to a person's heart is to talk about the things he or she treasures most."[3]

Being relevant—talking about things customers care about should be easy, especially with so much corporate focus on customer-focus. It isn't. In fact, marketers display a surprising degree of tone-deafness about this.

A *Harvard Business Review* article cited research done to understand why consumers interact with companies on social media. The study also asked marketers how they thought subjects would answer. In other words, investigators assessed marketers' knowledge of what consumers considered to be relevant. Pollsters asked 7,000 consumers across the U.S., U.K. and Australia to rank a dozen reasons, from most to least important, for why they interacted with companies on social media. Next, researchers asked Chief Marketing Officers and other marketing execs (200 of them, representing 125 brands in 12 industries) to sequence the same list in the order they believed consumers ranked them. Did marketer's mirror what consumers said? Not even close. In fact, marketers did not put a single reason in the same rank order as customers. Their answers were 100 percent wrong.[4]

Consumers cited "getting a discount" as the #1 reason. Marketers wrongly predicted consumers would rank discounts dead last. The #2 reason consumers cited was to purchase products, a reason ranked #11 by marketers. When it came to understanding what scored as relevant to consumers, marketers literally could not have been more wrong. There's a moral to this story for B2B marketers: get to know your customer.

Relevance: One concept, two advantages.

Being relevant can pay you back two ways:

1. It makes you interesting to others (customers, prospects, sellers, distributors, etc.). And people pay attention to what interests them.
2. Being more relevant than a competitor—makes you different. And isn't being different the whole point of differentiation?

Being relevant: "how to."

Carnegie tells us that to earn people's attention, interest, respect—and perhaps even their purchases—we should, "talk about the things he or she treasures most."

So how can we know what our audiences value most? Ask. But do you know how?

I pieced the following tips together from what we've learned in thousands of hours of asking questions and listening to customers and prospects of our clients. Understand we made plenty of mistakes along the way. Over time, we improved the methods, so that today, you can learn in a few minutes what took us 40-plus years to work out.

For a more complete discussion on how to gain insights into customers, I refer you to the appendix of this book and Chapter Three in *The People Powered Brand*, co-authored by me and my 5MetaCom colleague Joe Bannon.

A few tips about learning what's relevant to your audience . . .

- Get customer input from the source

  Be wary of customer opinions from anyone other than customers. This includes internal sources such as sales reps, technical support, or customer service people. These folks often have a good handle on customer issues, but personal bias—whether conscious or unconscious—invariably colors their views. Their own self-interests can rise to the surface. To get customer insights, interview customers. To get prospect insights, interview prospects. Sometimes you'll want salesperson, dealer, or distributor insights. When that's the case, don't guess. Interview those people.

- **Consider getting input from experts**

"An expert," said Nobel Prize–winning physicist Niels Bohr, "is a person who has found out by his own painful experience all the mistakes that one can make in a very narrow field." Agonizingly true. Industry consultants, university professors, trade publication editors/reporters and others, can all be expert sources. You may find, as we often do, that experts can provide you a more visionary and broad-based view than customers, prospects or employees. If you don't have the time or money for extensive discovery research, expert interviews can be a short cut to getting good and valid information quickly. In addition, we often find interviewing a few experts prior to talking to customers quickly gets us up to speed on relevant issues. It can make those customer conversations more productive.

In his book *How Customers Think,* Gerald Zaltman summarizes the work of researchers Robert Clemen and Robert Winkler, stating ". . . only about three experts who do not know each other are needed to identify the relevant issues in a problem area."[5]

Our experience supports their findings. But understand, it often takes some creativity to uncover a true expert. For instance, I once interviewed a trade publication editor who reported on the global feed manufacturing industry—the precise target market of our client. Prior to becoming a journalist, he spent years in the Peace Corps helping to design and build modern feed mills across Asia-Pacific, our target geographic market. As expected, this expert displayed an editor-in-chief's broad business perspective. We often find this with seasoned editors. Uniquely, however, he possessed a true "boots on the ground" perspective, having spent years inside manufacturing facilities working with the very management people we intended to target. The editor and I became friends and I turned to him often for insights to help our client. Perhaps it goes without saying, but finding the right expert can make all the difference in the depth and breadth of what you'll uncover.

- Recommended Format

For "insights" research, conduct one-on-one interviews of about 60 minutes in length, instead of quantitative surveys. Avoid focus groups if you can. The only advantage they provide over one-on-one interviews is cost-savings. In every other way, one-on-one is better. When your goal is to gain deep understanding, along with the "feel" and language of your customers, nothing beats in-depth interviews, one at a time. Use quantitative research (surveys) when you want statistical validity or the ability to project results. But when you want to know what your audiences "give a bleep" about, go with one-hour, one-on-one conversations.

- How many interviews?

MIT and the University of Chicago reported that twelve to fifteen 45–60-minute, one-on-one interviews with randomly selected subjects can reveal about 80 percent of unique "needs" of an audience, regardless of the size of market being studied.[6] Our experience agrees with this. In fact, after three or four interviews we usually start to hear a repeat of main themes. As we close in on a dozen, we're often able to predict how subjects will answer questions. That's good; it demonstrates a gained understanding of the customer.

- In-person, phone, virtual/web?

You'll gather more information and better information when you conduct interviews in-person, face-to-face in their world, where they work. You'll learn what's relevant when you can see things first-hand. Even more valuable is when they do some "show and tell." We once conducted a series of interviews with high profile architects at their offices. We traveled to New York, Los Angeles, Boston, Houston, and other major cities. In the research, we asked subjects to show us how they would get started solving a specific design problem. What we learned could not have been experienced remotely. Some of our subjects got up from their desk to pull items out of file drawers, closets, or bookcases. Several marched us down a hallway to grab a needed resource from an in-house library.

Curiously, not a single person reached for a phone, desktop, or laptop computer. These findings helped us create an all-new reference tool for architects and materials specifiers that better explained the utility of our client's product lines. The new tool made it faster and easier for architects to find a solution and to spec our client's products. Kudos to our client for approving in-field research versus cheaper phone or virtual interviews.

In cases where you don't have the necessary time or money for in-person work, connecting via video conference or phone still beats going without customer insights.

Questions: Start broad, then narrow in.

Don't start out your conversation by asking about your company or product. First, get to know the person and what they do. After opening pleasantries, start with a broad statement such as, "Tell me a little about you and what you do in your job." Or, "How'd you end up in this career?" Ask about big issues they face, trends they're seeing and how they're dealing with those.

I often ask, "What's a good day like around here?" And, "What's a bad day like?" On occasion, answers to those two questions alone have given us missing pieces of information we desperately wanted. Halfway or even two-thirds of the way through the interview is when to start narrowing in on specific issues.

Learn where your product's utility fits into their world. Are the issues you help with high on the list of things they care about? Or way down in the pack? When what you fix ranks high, you'll have more of an opportunity to command an audience's attention and ask for their time. When your issues rank low, you won't get much of either. All the more reason for you to stand out by being different, and to be brief and simple in what you have to say! More on that in Chapter Five on messaging.

For a list of 32 idea-starter questions that will work for most any interview in any market, go to the Appendix, page 123, or pages 47–48 in *The People Powered Brand.*

- It's a conversation.

Don't think of customer interviews as a survey or interrogation. I even hesitate to call it an interview. A very human and personal conversation is what you're after. Be genuine. Be curious. But beware; you can't fake it. People can sense when you truly care or if you're simply going through the motions. And how do you convey you care? By paying attention to what others say. By listening and reacting to what you hear. And by asking appropriate follow-on questions.

You should, of course, prepare a discussion guide—an outline of the topics to explore—in advance. At the same time, you should never hesitate to veer from it to explore off-script topics your subjects bring up. These can provide an opportunity to learn things you could never have dreamed of asking about.

Case in point: We once discovered an all-new use for a client's product in an interview. The newly discovered utility for the product ended up transforming the fortunes of my client and revolutionizing an industry, in the U.S. and around the world, all because I asked a series of follow-on questions to a casually uttered remark from a single customer. Listen. Empathize. React. Enjoy the learning. I consider interviews a form of conversational ping-pong. The more you listen, empathize and react with additional relevant questions, the more likely you'll learn things that spark new ideas for differentiating.

Summarize what you heard

I'm a former magazine writer and newspaper reporter. I take detailed interview notes. It helps me remember. It also signals to people I talk with how important their views are to me. But I do more than just take notes. After each interview, I immediately capture my top-of-mind observations by answering four questions:

1. What key ideas or themes emerged?
2. What did the subject care about the most?
3. What did I hear that surprised me?
4. What new issues came up that I should explore in later interviews?

For more on how to do this, refer to pages 50–53 in *The People Powered Brand.*

What about quantitative surveys?

These can be valuable, especially when you require statistical validity and projectability of results. But for getting to know how customers think and feel about things—to learn what's relevant and what stands out as different—nothing beats deep conversations, one at a time. In my search for differentiating ideas, in-depth qualitative interviews have never failed me. Although valuable in other ways, I can't make the same claim for quantitative research.

CHAPTER 4

Differentiating with a category

"First"

"Only"

"The Leader"

What marketer doesn't covet these positions?

Today, marketers, and nearly all businesspeople, understand the concepts of position and positioning. Few, however, know its origin. But they should.

In my experience, fewer still understand how effective positioning can differentiate products in the market to a level equal to *or beyond* what can be achieved from product innovation. In pursuit of being different, you'd do well to fully understand positioning, along with an allied concept: category. But first, a little history.

The late 1960s: New thinking on Positioning.

As a modern marketing concept, positioning arrived via a June 1969 article by Jack Trout in the trade publication *Industrial Marketing*. In 1972, Trout, teamed with Al Ries to co-write a three-part series for *Advertising Age* magazine that expanded on the idea. The pair, ad agency principals and alumni of General Electric's advertising and sales promotion department, laid out a now familiar lament discussed at the start of this book:

> "There are just too many products, too many companies, too much marketing 'noise.'"[1]

In article two, Trout and Ries continued along those lines:

"To cope with this assault on his or her mind, the average consumer has run out of brain power and mental ability.

". . . (the consumer) is less and less interested in making the 'best' choice.

". . . a 'satisfactory' brand is good enough."[2]

Sound familiar? And don't forget, they wrote this stuff in 1972!

In 1982, Trout and Ries authored the wildly successful book *Positioning: The Battle for Your Mind.* It's now in its fifth decade of publication. They urged marketers to use their new methods to fight for sales. They argued against a belief of the time that "positioning" was something done to the product. They argued against going to war for sales and share in the marketplace. Radically, they proposed taking the fight to a six-inch-wide battlefield, saying, "a company must create a 'position' in the prospect's mind." Marketing, for them, dealt with human behavior and psychology. Their concepts and examples dealt mainly with consumer goods. Application and enhancement of their ideas for technical and scientific B2B products fell on a small set of practitioners, including me and my 5MetaCom colleagues.

So why the history lesson on positioning? Because today we face the very same marketing challenges Trout and Ries encountered—complexity and noise—yet at a far greater scale. And so, the foundations of positioning—anchored in human psychology—still hold true for our present-day pursuit of differentiation.

Category: The Basics.

In *Positioning*, Trout and Ries talked about a companion idea: category. Our minds, they explained, use categories to organize all sorts of information, including products. To illustrate, they used the analogy of a ladder. Each ladder, they said, represented a category and each rung on a ladder belonged to a product brand. For example, on the "cola" ladder, Coke occupied the top rung, followed by Pepsi. The highest rung on every category ladder belonged to the top brand (almost always the leader in sales/market share). Competitors grappled in the consumer's mind for the category's top rung.

Before we get too far, let's review how categories and brand names work together. I'll use Red Bull® energy drink as an example. ***Red Bull*** is the product brand name and trademark. ***Energy drink*** is the category—the mental file folder in which products of this ilk reside.

Explained by Trout and Ries as a ladder with ranked brands, category seems simple enough. And in truth, it is. Perhaps that's why use of category as a marketing tool gets shrugged off in favor of more "sophisticated" —and often less effective—approaches. But some concepts remain true and valuable for the long haul. They outlast fads and even trends. I feel that way about category as a marketing idea. If I could select just one concept to put in my marketing toolbox, I would choose category. Used with strategic skill, category can unleash differentiation that transforms products and markets. Hyperbole? Not in my book.

Category: An organizational container *in the mind* into which we place similar things.

Instead of Trout and Ries' ladder analogy, I propose a folder. So, let's imagine one—a digital folder—the kind you use on your computer. We all know how this works. We give the folder a name, then place information (files) in it that fits the description. The folder name is a category.

On the other hand, our folder could be old-school, a paper one made of heavy card stock, with a notched tab. On it, we write a word, or several words to describe what's inside. The word(s) we write on the tab—that's a category.

Categories, whether digital or physical, give us a way to organize information. In our mind, categories do the same thing. *Brain Rules* author, neuroscientist Dr. John Medina, tells us our brain likes this type of order. And that's why, as a marketing tool, category holds so much inherent power. Category-style thinking comes as standard equipment in the mind of every one of our customers and prospects.

There's one more important point to know about categories. Despite what some marketers think, in the real world, people want, need, and *think* about categories long before they want, need, and think about brands.

Writing in *Ad Age,* Al Ries said nobody walks into a store and says, "I want to buy a Sony."[3] He's right. And the same idea applies to technical and scientific B2B goods.

I seriously doubt some lab manager out there is thinking, "I need a new BelArt® SP Scienceware® Riteflow® H404070035 SP."[4] Much more likely is, "We need a new mounted flowmeter."

Or consider the IT worker who tells a supervisor, "Boss, we need a new server." Notice the worker didn't say, "Boss, we need a new ProLiant DL580, Gen 10." Category first, then brand.

The minds of people who buy things—for their business and personal lives—operate like this all the time. For instance, it's nearly lunchtime as I write this. Know what? I'm thinking about a big, juicy hamburger, not a Wedgie™ Burger, SONIC® Crave, or a Thickburger.®[5–7] Sure, I may end up buying one of those, but only after I decide I want a hamburger for lunch. People think category first, then brand. That's why I say you must master this concept.

Prospects—people who can and should buy from you—will place your product in a category. And depending on which one, you may face massive competition or perhaps none at all. You could, for instance, be placed in a category that lumps you in with scores of others. In this case, you risk being thought of as a commodity, a "me too," even though your product may be quite different from others.

Ending up in the wrong category can literally make or break a pricing structure, product line, career, or company.

Get placed in a desirable category—one where your product stands alone, for example—and you'll set the benchmark for value and price. Being seen as the only product in a category also says customers and prospects recognize and buy into your uniqueness, your difference.

How to differentiate with a category

Perhaps you're thinking, "If categories deal with *similarity*, how can they *differentiate*?" Answer: when the category you're in contains only one product—yours. Achieve this and your product instantly receives title to being ***the first***, ***the only*** and ***the leader*** in the category.

And what if no category exists into which your product and yours alone can be placed? This is almost always the case. Simple: you invent one. This approach works particularly well for new products, but that's not a requirement. The new category approach can easily work for more mature products.

Creating a new, accurate, descriptive, differentiating *category* makes it easier for customers to understand the utility you deliver, making buying decisions simpler for your customers. The very best marketers center their efforts on helping the customer. And categories do just that. Categories help customers recognize and understand what's different between competing products. Marketers of course benefit when they get the category right; they sell more. But always remember, customers benefit first and most when an informative, differentiating category helps them get desired utility from a purchase.

Speaking of simplicity, the same *Harvard Business Review* article cited in Chapter Three highlighted its importance in making the sale. Authors Patrick Spenner and Karen Freeman in their article "To Keep Your Customers, Keep it Simple," stated the single biggest driver of consumer's likelihood on following through with an intended purchase, "by far, was decision simplicity—the ease with which consumers can gather trustworthy information about a product and confidently and efficiently weigh their purchase options. What consumers want from marketers is, simply, simplicity."[8]

Let's make sense of all this with a few B2B product marketing examples.

EXAMPLE #1
New category scorches competitors.

Background

Business insurers know when commercial buildings burn, water often causes more damage than the fire itself. The reason, of course, is that most buildings come equipped with sprinkler systems. When fire erupts, sprinklers kick in and douse the flames. As a result, commercial buildings rarely burn to the ground. But there's a catch. Water, spewing from fire-triggered sprinklers, can create untold havoc, drenching and destroying furnishings, equipment and infrastructure like computers and data storage that contain priceless customer and financial records, especially accounts receivable. Who knew life-saving water could cause so much destruction? I'll tell you who: our client.

They invented and marketed a system that put out fire without liquid of any kind. So, you might imagine their product would become #1 in their category, right? Wrong. That is until my colleagues and I invented something else for them: a new category for their product that left the water-based competition on the outside, looking in.

Think back to the file folder. Our client launched its technology into a category called "fire suppression." They didn't invent or willfully choose the category. It's just that everyone accepted and used that category name for products that put out fires in commercial buildings. The fire suppression folder contained scores of water-based products and systems, and our client's new product, as well. Not great for them.

Marketing-wise, the fire suppression category proved even worse than we could have imagined. We learned building owners mostly saw water as a good thing. After all, it put out fires, protected buildings and saved lives (all true, by the way). In fact, in the world of commercial buildings, water had become an accepted paradigm—*the* accepted paradigm. For architects, engineers, construction companies, and building owners, fire suppression meant water sprinklers and water.

Yet, here came an upstart, saying, "not so fast . . . use our product instead of water." But saying "our product is better than theirs" didn't resonate. Yes, it highlighted a difference (non-water), but it lacked relevance because non-water didn't fit the paradigm. This made it hard to gain attention and create meaning, and even harder to get considered. The tide worked against their non-water solution. Things remained this way until, ultimately, the non-water folks brought us on board.

Hypotheses

We started out with several hunches: 1) Their obstacles centered on a lack of relevant differentiation; and 2) Category could be a big part of the solution. After a series of discovery interviews, those notions proved true. And while the product was truly different, it had become anchored in a category with water. We were (forgive me) drowning in irrelevance.

So, we asked, "What if?"

What if we weren't lumped in with water? What if we created a different category? And what if the new category plainly pointed out our difference and hinted at problems caused by putting out fires with water? What if we turned the marketing fight from a clash of products, into a battle of categories: water versus non-water? Would people pay attention?

Solution

Step one: Create a new category. Step two: Focus our client's marketing on the differences between the two vastly different categories. With that, the new category we created, "Waterless Fire Protection," entered the fray.

As they say, "less is more." That one word—water*less*—turned the tide. It changed the conversation. It transformed stale product pitches into deep dialogues about category differences and utility. The generic benefit—putting out fires—became seen as a table stake. What the waterless category did not do became the differentiator.

And so, salespeople opened conversations like this, "I'm here to talk about fire protection for your buildings and the differences between sprinklers and *waterless* fire protection."

They talked about categories. Reactions from prospects tended to go like this, "Wait a minute . . . we use sprinklers." (Relevance.) And

"How can you put out fires without water? What's wrong with water?" (Differentiation.)

That one word—*waterless*—served double duty, making us relevant while spotlighting our difference. It implied problems with "the other guy." And it opened the door to discussions on the merit, utility, pros and cons of water versus water*less*. Sales conversations evolved naturally into dialogue about the ins and outs of our product line, installation procedures and economics. As is often the case, prospects in this market welcomed an exchange of ideas and opinions on a new and different category instead of a product pitch.

Did our client make every sale? Of course not. But did the game change? Yes. The marketplace accepted the new category. And de facto, our client's product became the leader of it—a happy ending, at least until their patent ran out. Sorry to bring you back to reality. But it's for good reason. My guess is you already know this: When you're successful differentiating, you'll inevitably invite and face competition, sometimes in the form of a generic. And that's why top companies and marketers are advised to constantly innovate with new products and differentiating marketing approaches. Always have something new and different in the works.

EXAMPLE #2
Category creates "best of both worlds" advantage.

Background

I vividly remember the scene. Six hours into the day-long differentiation workshop we had been facilitating, one of the participants (the inventor of the technology some 20 years earlier) abruptly rose from his chair. Jolting to the front of the room, he blurted out, "That's it." With right arm outstretched, he pointed and jabbed back-and-forth in the direction of the whiteboard. Half laughing, he bellowed, "That's it right there . . . right there . . . right there . . . right therrrrrrre!" And sure enough, it was. Decades after creating it, the father of the product had under our relentless prodding solved the vexing riddle of what made his creation different from all others. Ever the scientist, he continued, "I'll be honest. I never have trusted you marketing guys. Well . . . maybe I've been wrong about that. But not about the science . . . thank you!"

I sensed we had put the group through enough torture. "Right there," I echoed, pointing to the inventor's chemical structure diagram. "I see the foundation of a new category. You okay if we take it from here and develop it?"

In this example, a highly technical feature—at the molecular level—gave us proof, the "because" we needed to credibly devise a new scientifically sound marketing category and catapult sales to the top of their market.

Prior to the workshop, our client's product battled multiple competitors in a two-category marketplace. And for years, sales plodded.

To better illustrate, let's call the two categories A and B. Category A products contained carbon. Category B, the opposite: products without carbon. Our client's product contained no carbon, placing them squarely in B.

PhD-educated buyers considered Category A products the gold standard: high performance, high price. Category B cheap; not highly effective. For decades, buyers experimented, trying both A and B products. Some swore by A products. Others believed just as fervently in Category B products. In the end, many ended up blending the two to engineer good performance at a good price. And here's where things got complex.

Technically our client's product belonged in Category B. But it possessed an unusual feature, a result of their unique manufacturing process. This feature made their Category B product perform as well as A products. The more intricate (and costly) production method, however, meant pricing their B product higher than any other B competitor.

Third-party testing showed this B product performed like products in category A. Pricewise, our B sat mid-way between A and B. This odd combination just didn't make sense to buyers. Defined scientifically as a B product it couldn't shake the category's perception of low performance. Buyers struggled to accept a higher-priced B category product which they assumed didn't work particularly well.

Hypotheses

Our first inclination was to advise our client to go head-to-head with expensive, gold-standard A-category products. After all, their product performed similarly and had a nice price advantage. Background discovery research soon dashed those hopes. Skeptical PhD buyers couldn't get past the product's technical conundrum. They didn't accept that a non-carbon B-category product could perform as well as A-category products containing carbon. The science they knew meant our performance claims did not make sense or seem credible to them.

Solution

Faced with an "Innovator's Impasse" (the poem in Chapter One), we proposed a path forward: starting with a differentiation workshop.

Earlier, I told you how the workshop concluded, with a "eureka!" The whiteboard scribbles held the key to what happened next. Armed with the science, our team created a name for the new category based on the inventor's whiteboard molecular diagram—and a new brand name for the product. The category name we created contained a molecular, science-based root word. The brand name we devised hooked easily to the category. It hinted at how the molecule behaved to deliver "best of both worlds" cost and performance differences.

To launch the new category, our client's technical team put great effort into establishing the scientific validity of the category we created. They worked with leading institutions to publish peer-reviewed, technical papers that backed up the science. The new category became accepted and then entrenched. Today, a quick Google Scholar search for the category name we created shows thousands of citations. Not familiar with Google Scholar? As a B2B marketer, you should be. Check it out.

PR-wise, our campaign focused on the newly identified category. Branded marketing tactics centered on the product and its utility. In the end, boosted by a differentiation strategy of a new category, the Category C product became #1 across all competitors, eclipsing sales of products in A and B.

EXAMPLE #3
Category leadership feeds entrepreneurial hunger.

Background

This next example comes from David Karam, CEO of Sbarro Pizza—the world's #1 brand of impulse pizza. And what is "impulse" pizza? The kind that when you see and smell it, you "on impulse" stop what you're doing and buy a slice.

B2C Solution

Being omnipresent—constantly in front of consumers with well displayed, delicious looking and smelling pizza is key to Sbarro's success. You should hear David talk about the sensory lengths he and his store operators go to making Sbarro's product stand out—sparklingly-lit product display cases, stunning menu board photography, and plenty of ambiance that says, "hot, delicious, New York–style pizza . . . right here . . . right now!" Sbarro's marketing urges that little voice in your head to whisper, "C'mon, let's go for a slice of that great looking, great smelling pizza over there . . . right now!" And why wouldn't the little voice whisper that? Sbarro serves terrific food, quickly, and at a big quality step above typical fast food.

B2B Solution

There's another key to Sbarro marketing: the B2B sale. David Karam sells dreams—the opportunity to own, operate, and profit from a unique brand of pizza business. We're talking about Sbarro franchises.

David's great at selling impulse pizza to consumers. He's just as good at marketing Sbarro franchises to entrepreneurs. David knows these potential owners will, at some point, mention the Sbarro franchise idea to others: a spouse, friend, family member, business associate/partner, or lender. He recognizes this as the moment of truth—the instant that other person asks, "Sbarro Pizza? What? Why?" And that's when the category, impulse pizza, scores again. Prospective franchisee can confidently answer, "Sbarro? Maybe you've heard. They're #1 in 'impulse pizza.' And pizza is a 50-billion-dollar business in the U.S. I'm going with a market leader—Sbarro."

Industry website "franchise.com" referred to Sbarro Pizza as "the leading quick serve restaurant brand serving the *impulse pizza* category." Leader in the "impulse pizza" category? Yes! And who invented the now industry-accepted category? A savvy marketer who knows how to differentiate —David Karam.

EXAMPLE #4
What happens when you can't describe what you do? Nothing.

Background

We once invented category names for a client's technical consulting work. Before coming to us for help, this client called their offering "business consulting." Customers struggled to grasp the utility delivered by the nebulous category. Worse, the new division they created around this offering struggled to attract customers.

Solution

We created two new categories of service offerings for our client. The two category names—*optimatics* and *systematics*—set their story straight. We defined optimatics (from the word optimize) as: making existing systems and processes better. And systematics (from the word systematize) meant creating new systems and processes to increase efficiency and effectiveness.

From confusion to customer comprehension in two credibility-enhancing categories.

A few tips about categories . . .

- Ground Rules

  Categories can and should never be brand names or trademarks, and vice versa. Brand names cannot be categories. Every product belongs in a named category. And every product has a name. More about names in Chapter Six.

- **Categories can define things other than products.**
  What else, besides your product, would benefit from a category name? We once helped a client whose product was so technically complex that until prospects understood the science, they couldn't possibly understand how and why their products worked or the utility they delivered. To solve the problem, we created a category name for the science our client's R&D division conducted. The new category gave them a way to explain their product innovations. This paved the way for customers to understand what made the products' offerings different and why those products provided unique utility.

- **Credibility is key!**
  A category name must be technically credible. Otherwise, it will be seen as a loosely veiled attempt at marketing. With categories, even a whiff of marketing spin will cost you credibility and spell doom. Avoid words that might seem cute, clever, "slogan-ish" or even hint at being "sales-y." In most cases, your own internal technical people will be able to warn you when you've gone too far in that direction.

- **Tell your story**
  If not completely intuitive, the category needs a story, an explanation that makes intuitive sense. In the earlier fire suppression example, waterless perfectly described what is meant by the category name—no explanation required. When you're successful with your story, those in your industry, both technical and non-technical people, will accept that your new category exists and recognize it to be uniquely different from other categories.

- **How to know when you need a new category?**
  Don't overthink this one. You'll know a bad fit when you see it. It's when your product gets lumped in with others that are not at all the same as yours.

New category advantages

When you create a new category, your product instantly becomes: "the first," "the only" and the "leader." Done right, the new category you create will be viewed as interesting and different. And when you invent a category, you establish yourself as an innovator. This proves your credentials and earns you credibility. Mercedes, for years, stated in its TV advertising, "We invented the automobile." That fact conveys knowledge. It establishes credibility more so than had they said, "We know automobiles." Or, "We make the best automobiles."

Category "leader" means #1

Just to be clear, leader represents an idea. You do not necessarily need to use that exact word. For instance, you can reference market share and say, "#1 brand of (***category***)." You can cite customer buying behavior: "Preferred 3 to 1." Or, "Installed on more offshore oil platforms than any other telemetry system."

You can be the leader in gaining new customers by saying "fastest growing." How you communicate category leadership is a strategic and creative choice.

Don't miss the point. Being a leader, or #1, or most preferred gives you a giant point of differentiation. I'll paraphrase Al Ries by saying the most important difference about the category leader is . . . *they are the leader.*

In *Influence: The Psychology of Persuasion*, Dr. Robert Cialdini writes at length about how greatly people care and can be swayed by how others think and behave.

Are you a leader? Say so. If you are not the leader, take another look. Dig deeper. Surely there's something you lead in!

Timing

It's best to launch a new category at the same time you launch a new product. But this is not absolutely necessary. In several cases, our agency has been able to successfully create a new category years after a product launched in its original "bad fit" category.

- **It's not (always) about you.**
  A new category can, of course, define you. Yet, a skillfully constructed category can also sometimes pigeonhole and constrain your competitor, as was the case in the waterless fire suppression example. This new category made water the bad guy.

- **Categories make news.**
  Categories are noteworthy and *news*worthy. As marketers, we want to talk about our new products. But customers and the media usually find new categories more intriguing. For them, a new category *is* news. Your industry's media will likely publish stories about a new category. For instance, an announcement about an all-new type of battery (with a new name for the new category) will receive greater coverage and social buzz than a press release about a new model of an electric car. Launching a new category and product? Take advantage of the PR opportunity. Share the category news first. When you do a good job with your category strategy and name, interest in your new product will naturally follow.

- **Patience . . . patience.**
  I read recently that it took a full 25 years for the personal computer category to reach peak sales. That's a long time for such a revolutionary and important innovation and category, and a good reminder that important, successful endeavors can take time. Be patient and enjoy the ride.

CHAPTER 5

Differentiating with the message

Pop quiz:

Q1: What is a *marketing message*?

A: A summary of your product's *difference, value,* and *reason to believe.* It is a marketing communications résumé for your product.

Résumé: *a brief written account of professional qualifications and experience.*

Q2: What's the *most important thing* a marketing message can and should do? (C'mon, you know this . . . re-read Chapter One if you must!)

A: Yes . . . *communicate your difference!*

Why? **Because you cannot sell anything when people ignore you.** Standing out, being different, makes it extremely hard for people to ignore you.

So that's it? Gain attention and game over. We make the sale? Perhaps, if you sell candy bars. But when you market technical B2B products, gaining attention simply begins the selling process. There's much more to making a technical sale than simply getting attention. But grabbing attention always represents the first step. It must occur.

There's also more to the marketing message than simply gaining attention. But gaining attention is the most important and first task it must accomplish.

Q3: What else makes for an effective marketing message?

A: Three qualities: *brevity, clarity,* and *impact.*

Marketing Message = strategy (not ad copy!)

Let's clear something up. A Marketing Message is about strategy. It distills and articulates the key points you will use to differentiate and make your product relevant. It's not an ad. It's not website copy, a slogan, or anything else placed in front of a customer.

If this is true, then why do we need a Marketing Message? To provide guidance. Think of it as a gyroscope. Its job is to keep marketing communications on course. A Marketing Message instructs those responsible for the product's success how it will be portrayed—internally and externally.

As a marketer, you'll of course communicate about your product externally to customers, prospects, resellers, and others. In addition, you'll want many inside the organization to know how you're portraying the product in the market. This could include senior leadership, R&D people, technical, sales, customer service, manufacturing, logistics, key suppliers, and others.

As a résumé of the product's professional qualifications, the Marketing Message, briefly lists what type of work it can do. It goes through the promises it makes and also includes references to back up those claims. It puts into words the strategic intent of ***what*** to say about the product. It should also guide development of the creative marketing communication work that will follow. In the next chapter we'll deal with creativity: ***how*** to say your message.

What goes into the Marketing Message?

Answers to three questions about the product:

1. **What's different?**
   Purpose: to capture *attention*
2. **What's worthwhile and valuable?**
   Purpose: to express *relevance* and *utility*
3. **Why believe?**
   Purpose: to establish *credibility*

Marketing Message—general approach

The method I'm about to explain stands on a solid footing: the psychology of how people's brains work.

Consider this from neuroscientist Dr. John Medina in *Brain Rules*:

> "Introductions are everything. If you are trying to get information across to someone, a compelling introduction may be the most important single factor in the success of your mission."

And this:

> "The brain processes meaning before detail, and the brain likes hierarchy. Starting with general concepts naturally leads to explaining information in a hierarchical fashion. Give the general idea first, before diving into details, and you will see a 40-percent improvement in understanding."[1]

The Introduction—for brevity and impact

Given the importance of the introduction, here's my advice:

1. Make sure it includes your *difference*. That's how you capture attention.
2. Keep it short. You have only a few seconds to become relevant and make an impact.

Much of the book up until now has dealt with item #1. Now let's go deeper on #2 to examine the "on-off" nature of the human attention span which demonstrates the need for brevity and relevance in your introduction.

In *The Advertised Mind*, market researcher Eric du Plessis reports on an investigation by H.E. Krugman that measured attention spans for advertising. According to du Plessis, Krugman's study showed: "the viewer's attention in the first 4-to-10 seconds of a television commercial shows with 83-percent accuracy whether he or she will give it attention over the entire 60 seconds."[2] (The experiment used 60-second spots.) In other words, audiences grant you mere seconds to capture their attention and interest. Do so and they stay with you. Fail, and they *quickly* tune you out. With apologies to Shakespeare, *brevity* is the soul of . . . an effective introduction.

As for how long a person must pay attention before your message achieves impact, I'll refer to empirical data contained in my earlier book, *Brand Busters.* It cites advertising recall and recognition research showing that ads looked at for longer than 2.75 seconds broke through, made a lasting impact, and were later recalled; anything less and it was as if the message had never been seen.[3] This research showed when you fail to spark interest in the first three seconds, people tune you out.

> **Conclusion:** to achieve results, the introduction to your marketing message must be short, no longer than 10 seconds, and not less than about three seconds.

Could the introduction be longer? I suppose so. But the data show it shouldn't be. Shorter than three seconds? I don't see how. Just know this: pack your absolute best stuff into the first 10 seconds of your Marketing Message. Will this be a struggle? Probably. But why violate these guidelines? Be disciplined and be brief and you'll skew the odds in your favor.

The rest of the Marketing Message—the rule of three for clarity and believability

Effective messaging for complex products requires more than just a short, albeit impactful or even emotional introduction. Since we're dealing with B2B considered purchases, this requires additional rational content to express value and also provide reasons to believe stated claims. When you do this, you'll achieve the other two goals for an effective Marketing Message: expressing *utility* and achieving *credibility.*

Medina says to communicate successfully requires hierarchy and organization. But how? Several psychological experiments provide instruction on how to do this and also boost credibility at the same time.

Do any of these phrases sound familiar?

- Life, liberty and the pursuit of happiness;
- Gold, silver, bronze;
- It's as simple as 1, 2, 3.

My guess is you know them. The first comes from the U.S. Declaration of Independence. Next, the three medals of the Olympics—awarded since ancient times. And who hasn't heard countless instructions and marketing pitches delivered in a 1, 2, 3 format?

These all follow a three-element organizational structure. From "three blind mice" to the kid's fire safety, motto of "stop, drop and roll," this approach has been with us since before we could read. The late Apple CEO Steve Jobs used it to introduce the world-changing iPhone in 2007, calling it, "Three revolutionary products: an iPod, a phone, and an internet communicator." He unveiled iPad2 in 2011 as, "A smaller version of the iPad that's thinner, lighter, and faster." And while these products were mostly consumer goods and not B2B, they were innovations of a complex, technical nature. Jobs did not let that stand in the way of brevity, clarity, and impact in his messaging.

Most of us would agree the pattern of three makes things easy to understand and remember. It seems to produce brevity, clarity and impact as well. But is this true? Are there data . . . proof? Does a "rule of three" really exist? The answer: yes.

Business professors at Georgetown University and UCLA investigated the persuasive effect of varying numbers of marketing claims made for products. They published their findings in a 2014 *Journal of Marketing* article: "When Three Charms but Four Alarms: Identifying the Optimal Number of Claims in Persuasion Settings." The paper details their experiments, conducted for four products which showed that when it comes to persuasion, more claims are better—until the fourth claim.

After introducing the studied item, they measured attitudes toward it and found positive feelings increased as claims went from one to two and from two to three. (They studied one to six claims.) Their experiments also revealed skepticism remained stable for the first three claims and then rose with each additional claim thereafter. To enhance persuasion and minimize skepticism, their studies point to three as the optimal number of claims to make.[4]

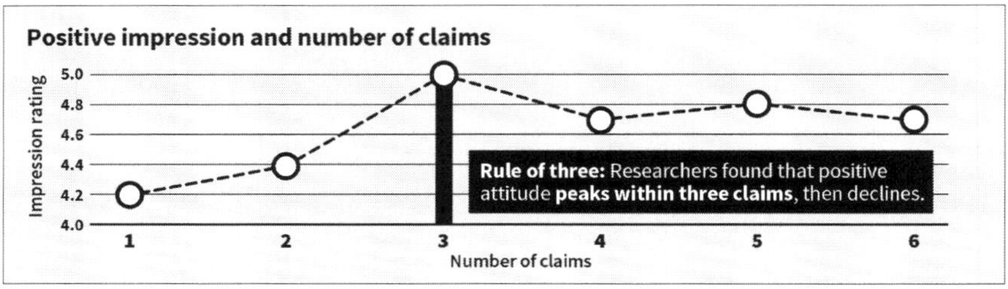

One experiment examined the relationship between the number of claims made and the resulting impressions of a product or person. The study found three claims is optimal.

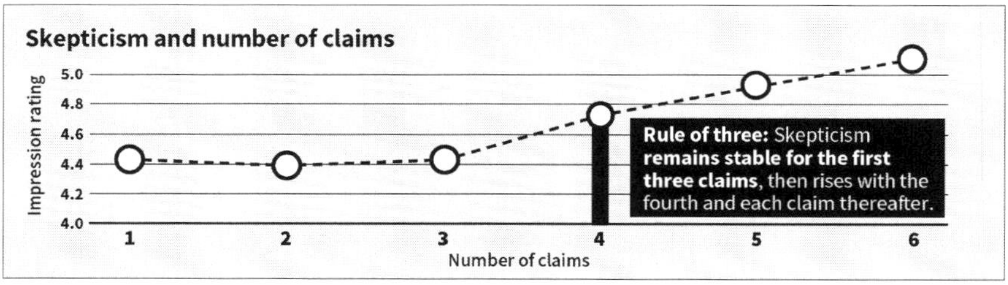

A second experiment gauged attitude and skepticism in relation to number of claims. It found that attitudes are most positive up to three claims, and that skepticism increases for each claim beyond three.

What about memory? We certainly want audiences to remember and recall our messages. How many "chunks" of information can people retain and recall? In other words, when we make three claims, can they, will they, remember? The answer: yes.

> "The preponderance of evidence . . . strongly suggests a mean memory capacity in adults of three to five chunks, whereas individual scores appear to range more widely from about two up to six chunks."

That comment concludes a paper from Dr. Nelson Cowan, Professor of Psychology, University of Wisconsin–Madison. In his paper, "The Magical Number 4 in Short Term Memory," Cowan describes his review of existing literature on memory (citing 439 references), in which he sought to settle the scientific score on how many "chunks" (his term) of information people could remember. As the paper title suggests, he landed on four as the "magical number," with some leeway (two to six) to account for individual differences.[5]

Among the many references Cowan cites, one from D.E. Broadbent indicated ". . . the number of items that can be recalled reliably, with very high accuracy, is about three or four . . . "

The evidence from multiple sources speaks loudly, with this conclusion:

> For maximum effect, marketers should organize and support their introduction (main message) around three main points/claims. Doing so improves persuasion and memory retention and keeps skepticism low.

Wait a minute! An introduction, plus three supporting claims—that's four things, right? Doesn't this ignore the rule of three? No. The introduction is not a proof point/claim. And don't forget, research tells us people can remember four chunks, an introduction and three claims.

Tools: an approach for your Marketing Message

Summarizing what's been discussed so far, an effective Marketing Message consists of:

- **An introductory statement**—the "main message," a brief "headline," 3–10 seconds in length, that captures attention and highlights the product's *difference.*
- **Three support messages**—that express the product's value and utility and establish credibility.

Neuroscientist John Medina gives us context for this line of attack:

> "If you want people to be able to pay attention, don't start with details. Start with the key ideas and, in a hierarchical fashion, form the details around these larger notions. Meaning before details."[6]

Before I introduce a template you can use for your Marketing Message, let's look at the rationale for it. Earlier, I likened the Marketing Message to a résumé of your product's professional qualifications. Most B2B products I've worked with come backed by lots of supporting information and credentials. Consequently, it's all too easy to follow a crisp, effective introductory headline with an overload of poorly organized support information.

In our work with marketing and sales organizations large and small, in the U.S. and around the world, we find most sellers, when given enough time, have little problem explaining the ins and outs of a product. Give a salesperson 30 minutes to talk about a product and they'll cover all the important information . . . *eventually*. But most times, they won't present it *properly*. They won't start with an attention-capturing headline. They won't follow that with a summary of key ideas and then form additional details around these larger notions. An unfocused, rambling dissertation like this could hardly be more out of step with our "limited time, unlimited choices" world.

To avoid this problem, we use and teach a structure we call 3-30-3-30. The method provides a way to organize and deliver a marketing message with attention-grabbing impact and successively greater degrees of completeness.

With the 3-30-3-30 approach, you can match messages to specific marketing situations and opportunities, whether short or long, without compromising brevity, clarity, impact, or completeness. It gives you a message strategy for everyday marketing settings—from a 30-second online video or "elevator pitch" to a 30-minute webinar or in-person presentation, and every type of product promotion in-between. Later I'll show you an example, but first, let's go over how the method works.

The 3-30-3-30 method

- **The first "3" stands for the 3-second *headline* (:03)**—an introductory statement to capture attention. It should be around 3 seconds and up to 10.
- **The "30" represents 30 seconds (:30) for *support* messages**—headlines for your three support messages. One roughly 10 second headline for each of your three support message = 30 seconds.
- If around 30 seconds is all the time you're given, use the :03-:30. Recall that Steve Jobs used this very approach to introduce the iPhone. "Three products: an iPod, a phone, and a revolutionary internet communicator." And he did it in less than 33 seconds. But we all know, you can't deliver a full set of professional qualifications, the product résumé, in a mere 33 seconds. Many marketing situations demand and allow for more and that's what the final 3-30 does.
- **The next "3" stands for three minutes (3:00) for *proof points***—additional depth added to the :33 top level content, achieved by placing three points under each support message (nine in total). These add substantiation (proof) to your support message. Cover three proof points for each support message and you're now at roughly three minutes.
- **The final "30" represents 30 minutes (30:00), the complete marketing résumé.** Here you fill in detail appropriate for the situation and audience, arranged within the coherent hierarchy set forth by the :03-:30-3:00 structure. When the situation allows for more time, an hour or two or even a full day seminar, simply add more detail under an appropriate support message.

  Example: after his :03-:30 iPhone introduction, Steve Jobs continued his talk using the organizational hierarchy he began with. In succession, he explained the value and "reason to believe" for each of the iPhone's three products. Jobs didn't invent the 3-30-3-30 format, nor did I. Some suggest it may have originated in the world of Washington politics, a culture driven by brief, clear, and impactful sound bites.

Tools—Marketing Message template

To simplify organization of your Marketing Message, 5MetaCom created a fill-in-the-blank template. We use it in our work with clients.

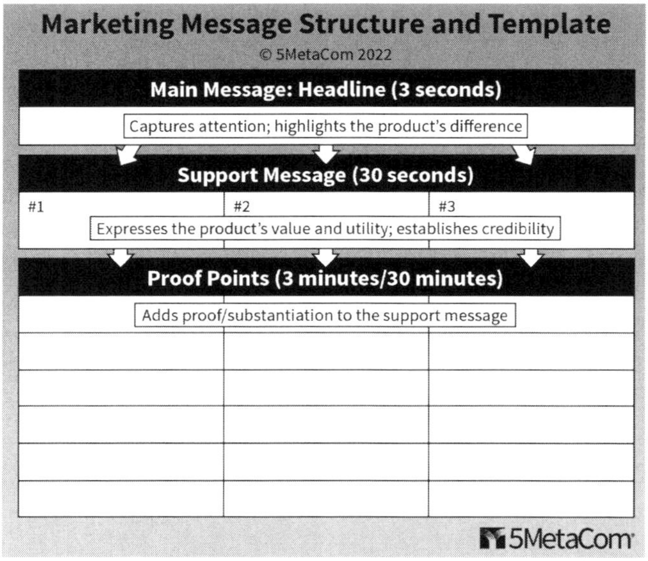

Developing the content

While the Marketing Message structure may be self-explanatory, developing the content is anything but. How you populate the grid is up to you. A single person can develop the content and then ask a cross-functional team to review (and hopefully improve) it. As an option, the content could be built-out by a team in a facilitator-led workshop.

Here are some general suggestions for content and the process for your Marketing Message:

What to include

- As discussed earlier, the content must answer three questions:
    1. What's different?
    2. What's worthwhile and valuable?
    3. Why believe?

- Review the key concepts from Chapters One, Two, Three and Four. Let those ideas inspire and inform your answers to the three questions. When you do, you'll pack your Marketing Message with power: ideas that differentiate, speak with relevance, capture attention, persuade, and establish credibility.
- For a unique perspective on how to come up with the right stuff, I refer you to the classic work: *A Technique for Producing Ideas* by James Webb Young. Written in the 1940s, this short masterpiece (roughly 40 pages) remains in print. In my experience, those who read Young's handbook and embrace its concepts quickly improve their ability to produce viable ideas. Have you read it? If so, re-read it. If not, will you?

The process

- Start by creating the :03 Headline and three Support Messages that make up the :30. This is the briefest condensation of your full Marketing Message. We call it the "3-30."
- From there, develop three Proof Points, in bullet point form, for each Support Message. You'll end up with nine Proof Point bullets in total. This yields about three minutes of information, the :03-:30-3:00. Bullet points, please. This isn't finished ad copy.
- Create additional sub-point bullets as warranted under each Proof Point, as many as you need to deliver the full Marketing Résumé for the product. If needed, use a second page.
- At the end of the process, you'll find yourself with a half hour or more of information in bullet point, summary form.

Example Marketing Message in finished form

Initially, I considered introducing the following example with a stage-setting marketing rationale. But things don't work that way in the real world. A prospect or customer doesn't care about your internally-focused rationale. In the real world, you get no opportunity for a pre-message business case discussion.

Below, you'll see a fictitious example about an industrial lift, a product used on high-rise building construction job-sites. This category of product exists, but the example is fiction—features, benefits, and even the product name, I made them up. *Do* pay attention to the hierarchy and format. *Don't* get hung up on details. I left many of them out. But be aware that in a real world assignment, extra details would be included.

Ready? I've numbered the sections; read through them in that order.

1 ProSafe Speed Lift—#1 for U.S. high-rise construction sites		
2 Safety— fewest accidents and claims; #1 in job-site safety	**3 Efficiency— the only 4 ton lift that hauls materials and personnel >150 ft./min.**	**4 America's #1 choice —more high-rise installations than any other lift**

5
• **Built-in safety features reduce 3 most common job site accidents**
  • Cite claim-rate data on the 3 accidents
• **Just two control switches—simple to operate safely**
  • Step-by-step operating procedure
• **Fully enclosed car protects workers in transit and others on-site**
  • Secures workers, tools and materials
  • Car design safety-feature details
	6
	• **Lift 40 workers or 4 tons of material 300 feet in less than 2 minutes with near zero downtime**
	  • Specifications: speed, car sizes, capacities
	  • 99+% uptime performance data
	• **No crane required for installation and tear-down: fast, cost-efficient**
	  • Install/tear-down details and features
	• **Lift heights from 100 to 500 feet**
	  • Details, specifications, models; utility for specific building types
		7
		• **Preferred 2 to 1 over other lifts for 10+ story construction sites across U.S.**
		  • Installations in all 50 states
		• **600+ installations—commercial and residential—since introduction**
		  • Cite data on locations by type of buildings
		• **Nationwide distribution and technical network**
		  • Inventory and service nearby, wherever you build in the U.S. (cite relevant locations)

Okay, perhaps I overdid it. Neither you nor I may ever have the pleasure of working on such an exceptional product. Regardless, notice how this Marketing Message delivers on differentiation, value/utility, and reasons to believe. Hopefully, it also checks boxes for brevity, clarity and impact.

Here's what I want you to take away from the example:

- **The :03 Headline (❶)** delivers facts that quickly introduce what makes the product different: it's #1, and there's always and only one #1. The product name also contains several important emotional features (safety and speed) that suggest a *because* for the product's leadership.
- **The :30 Support Messages (❷ ❸ ❹)** follow the organizational hierarchy established by the headline. As the headline talks about safety, efficiency, and the fact it's #1, so do the Support Messages. Each support message makes up about 10 seconds. When coupled to the headline, the resulting 33 seconds give the audience a condensed version of the complete ProSafe Speed Lift story.
- **Nine Proof Points deliver about 3:00 of information (❺ ❻ ❼)**, three for each Support Message. The pattern continues to tap into the power of three.
- **Sub-bullets under each Proof Point complete the message.** In my example, I went light on these and used broad ideas. You'll want to include specifics and additional sub-bullets as appropriate for your audience and the situation.

Marketers are (wisely) putting more time and effort into their messages.

Successful marketing involves a group of well-designed and executed pieces and parts. Among the most important is the Marketing Message: *what* to say about the product. In the past few years, I've noticed a heightened interest from marketers in getting their Marketing Message right. "We need help with what to say about our product," is the #1 request we now get from clients.

The Marketing Message method and format I've walked you through has been used for numerous products marketed in the U.S. and around the world. Clients tell us it works. As a tool, it can help you gain attention and differentiate in the "limited time/unlimited choices" world your audiences find themselves caught in.

A few tips about your Marketing Message . . .

- So, just drop some bullets into the grid and you've automatically got an effective message?
  Not at all. Recall that you must be different and relevant. You must say things people care about. The template is simply the canvas onto which you "paint" your message. It provides an organizational format. You and I know a quality canvas doesn't inevitably produce a good painting. In the same way, when you put bad content into the template, you end up with a lousy, albeit nicely organized Marketing Message.

- Remember: It's not ad copy!
  Therefore, don't massage the words in the Marketing Message to death. There's a point where the bulleted *ideas* in the grid are just fine "as is." Once you get there, stop. It's not finished promotional copy. The Marketing Message explains ***what*** to say when selling, marketing, and communicating. ***How*** to say it—that's where creativity should come in to add interest and power to the message, without altering its content! (The next chapter covers creative.) Apologies for all the exclamation points! Forgive me, but I simply cannot tell you how many times I've seen the pursuit of perfect bullet copy in this template ruin otherwise fabulous ideas and strategy. Don't do it!!!

- The Marketing Message is a guide (not a script).
  Its intended use is as a content guide for people who create and deliver messages about the product. This includes company salespeople, distributors, dealers and resellers, technical support folks,

creative teams (writers, designers, animators, videographers, etc.), and all others involved in selling and marketing the product.

We often train sales organizations on how to use the 3-30-3-30 approach in their interactions with customers and prospects. One thing we tell them not to do is memorize and recite the Marketing Message verbatim. The message, as contained in the 3-30-3-30 format is a guide for what to say. We encourage salespeople to convey the points contained in the Marketing Message in their own way. We encourage them not to change the hierarchy and content, but rather, to convey it genuinely and personally. I tell salespeople, "Say this, but put it in your own words." No customer wants to hear a memorized speech from a salesperson.

Get your people on-board

From an execution viewpoint, the most successful organizations "enroll" their people in the Marketing Message—explaining its purpose and rationale. The best way to accomplish this is in a half-day, interactive workshop. For a detailed look at how to do this, refer to Chapter Six in our earlier book, *The People Powered Brand.*

The power of admitting a negative

Business-to-business purchases often come down to balancing risk and reward. Top marketing and salespeople know this. They work hard to earn trust and reduce a purchaser's fear of risk. One powerful way to do this is by admitting negatives—sharing pros and cons about a product. Numerous psychological experiments have shown that positives result from admitting a negative. In one such study, participants were asked to read a confusing paragraph. Some readers were given the paragraph preceded by a written statement acknowledging the unclear nature of the text. Others were provided only the confusing paragraph. Participants who read the negative acknowledgment rated the text as clearer than did those who read the paragraph without the introduction, even though the paragraph given to both groups was identical.[7] Other research has shown presenting both sides of an argument—the pros and cons—to be more persuasive than arguments that present only positives.

I once tested ad copy for a product that included a statement about how the product would not make a big difference in the buyer's business results. The copy modestly stated the product would produce a small but measurable improvement, a truthful, albeit somewhat negative comment. That statement received more "like" comments than anything else in the ad—more than all the other words, photos, logos and graphics. My client hesitated to include such a negative statement in their marketing. I somehow persuaded them to let it stand for the ad test, if only to prove me wrong.

In the end, we learned how powerful the statement turned out to be. That very ad campaign kicked off a decade of year-on-year growth for what had become a sleepy, "me too" product. Sales in that timeframe increased from around $20 million to more than $100 million. This success came from much more than just the advertising. However, admitting a negative turned out to set the right tone with the audience. Compared to the competition, this was different. Ultimately, candor contributed to earning buyers' trust.

- Message Testing: How to know if your message is any good.

Message testing is a good idea, especially when there's no internal consensus on what to include or exclude. Our agency has developed a message testing methodology that reveals what an audience likes, dislikes, and finds confusing about a message. The technique can also be used to evaluate websites, packaging, sales materials, and other types of marketing tactics. Getting feedback—internally and externally, even if only informally—is always good practice. If message testing isn't your area of expertise, there are plenty of pros out there who can do this for you.

- **Remember: The main mission of marketing is to get new customers.**

  Therefore, your Marketing Message should be oriented to this. Other tools, services and efforts help *keep* customers: technical and customer service, education and training, financing, you name it. But first and foremost, your marketing and your Marketing Message should primarily be directed to prospective customers.

- **Confusion = death of interest and comprehension**

  Forget about being witty, coy or mysterious in your Marketing Message (and your creative work). Novices often believe cleverness captures attention. It does not. It turns readers off and makes them flee. I'll cite Dr. John Medina's neuroscience work again. "Normally," he says, "if we don't know the gist—the meaning—of information, we are unlikely to pay attention to its details. The brain selects meaning-laden information for further processing and leaves the rest alone."[8]

  John Hallward offers similar advice based on his decades of TV ad testing work: "It is a mistake to assume that an audience will put efforts into figuring out what a commercial is about," he says. "Usually they will not; if they feel even slightly confused, they just ignore the commercial."[9]

- **Is "Marketing Message" the same as a "Value Proposition?"**

  While similar, the two differ in length and detail. The Value Proposition consists of just the :03 Headline and the :30 message (the three Support Messages). That's our definition of Value Proposition. Do a little Googling and you'll likely find scores of different models, formats and definitions for Value Proposition. Feel free to use ours. It keeps things simple and consistent with the Marketing Message.

CHAPTER 6

Differentiating with creativity

Effective differentiation requires creativity—in strategy and message—but especially so in execution. I'm talking about all the tools and tactics you put into the marketplace to bring your strategy and message to life. This includes advertising, websites, email, social, video campaigns, sales literature, and support materials.

Sad to say, but a brilliant strategy and message, when executed in bland, boring, irrelevant creative work, can never hope to engage the interest of B2B buyers. It can't. It won't. Creative work must bring your difference, your relevance, and your utility to life, or it stands little chance of producing good results. And as a differentiation disciple, you know why.

Think back to the words of Jack Trout and Al Ries from Chapter Four:

> ". . . too many products, too many companies, too much marketing 'noise.'"

Yes! And that's why this marketing formula I devised holds true:

> Great strategy + great message + lousy creative work = an ignorable, irrelevant marketing campaign, which in turn = lousy, disappointing, suboptimized results.

Given the importance of creative work, we should see lots of striking, captivating B2B creative work in the marketplace. But we don't. Instead, we see lots of sameness. And sameness is the exact opposite of differentiation! How can any marketer hope to differentiate with sameness of creative work?

For fun, I Googled "sameness in B2B advertising" and ran into an article in *Forbes* online: In the article, "Message to B2B Marketers: We Can Be Creative," the author asks:

> "When was the last time you were inspired by a B2B marketing campaign? Chances are, not recently."[1]

I agree.

Lots of money; few ideas.

In our wondrous and abundant modern world, some things, stubbornly, remain scarce. Swivel your head and take a look around at the developed world and you'll see money isn't one of them. No, the scarce item I'm talking about is creative *ideas*.

But why so scarce? After all, isn't creativity and the *ideas* it hatches the solution to nearly every business or marketing problem we face? And if so, how do we explain the lack of time, effort, training, and business resources devoted to teaching, learning and studying, and continuous improvement in creativity?

Creativity: Doodling or serious business?

At the root of skepticism about the value of creativity may be its kinship with fine arts (music, painting, poetry, etc.) and entertainment (comedy, film, social media and the like).

Compared to the "real" world of business, creativity can seem frivolous, trifling. Unlike exacting fields such as engineering, accounting, and manufacturing, "creative work" seems wholly made up—random, you might say. And indeed, it is. Maybe that's why some businesspeople in the B2B world consider creativity a lesser pursuit, or at the very least, something that's hard to place a value on.

Yet isn't it interesting that in R&D, for instance, the ability to "make things up" gets held in high regard? Everyone knows product innovation, or lack thereof, can make or break an enterprise. But how about marketing innovation, otherwise known as creativity? Can it contribute equally, or perhaps more, to the organization's success or failure? I say yes. And that's why I also believe creativity in business deserves more emphasis. Looking specifically at advertising, here are a few things we know about the business value of marketing creativity:

> "Creative advertising was found to enhance recognition, and this positive effect increased over time."—Baack, Wilson and Till, 2008.[2]

> " . . . original advertisements drew more attention to the advertised brand." (than non-original ads) —Pieters, Warlop and Wedel, 2002.[3]

> "The use of a brand-differentiating message and a strong product focus continue to manifest a positive impact on measures of recall, comprehension and persuasion."—Stewart and Koslow, 1989.[4]

Drawing more attention, enhanced recognition, better recall, comprehension and persuasion for a product. Who doesn't want these positive business results?

Please don't misunderstand. I'm not trying to convince you to spend money on marketing and advertising, although in many cases, I believe you should. If you're a believer, I'm urging you to pursue creativity, excellence and originality to differentiate and thereby gain an advantage.

What makes advertising and marketing "creative?"

Creativity means expressing a Marketing Message in ways that capture attention and appear meaningful and different to your audience. A couple of marketing professors explain:

> "The first and most fundamental characteristic of ad creativity is divergence—the ad must contain elements that are novel, different or unusual in some way."—Yang and Smith, 2004.[5]

Continuing, they say, "creative ads are those perceived to be divergent *and* relevant." Hooray! These concepts, divergent (different) and relevant, should be quite familiar to you by now.

Creativity merges difference and relevance into communication that moves an audience to act, to *do* something: pay attention, keep reading, think deeply, watch a video, attend a webinar, request a demo, contact a salesperson, etc. What's it worth to your company when a prospect takes one of these actions? With a little work, you should be able to quantify the value of marketing creativity to your business. Try it. Then write a book about it!

The Great Debate: rational vs. emotional appeals in B2B marketing.

Do B2B buyers make purely rational purchase decisions?* Or, does emotion rule? Do their psychological purchasing styles flip flop between rational and emotional? Or in the buyers' mind, does some combination of emotion and logic ultimately decide?

These questions form the crux of one of marketing's great debates. Which appeal—rational or emotional—should we use to maximize results?

You're about to learn the answer . . .

Which appeal to use?

Six months ago, I set out to research this question and settle the debate. The good news: I found plenty of consumer psychology advertising experiments on the topic. The not so good news: most of them—all of them, really—didn't precisely fit with B2B.

Some studies showed which appeal worked best in a given marketing situation: mature vs. developing markets; a highly technical buyer (geek) vs. a generalist; a relevant product vs. one for which relevancy must be established. Others indicated which appeal to use depending on goals: awareness building vs. moving the brand into the consideration set vs. stimulating purchase vs. getting the buyer to understand the product more deeply. Others demonstrated which type of product would benefit most from a particular appeal: highly differentiated vs. "me too" goods vs. impulse purchases vs. considered purchase/durable goods.

---

* Since the early 1900s, economists have referred to a mankind that functions in this manner as *homo economicus,* or "economic human," a species that makes purely rational choices based on economic self-interest.[6]

Along with the truth, I had hoped to find an easy answer. But as with most things in business, forget "easy." And the truth ended up being, "It depends."

Slowed, but not defeated, I analyzed the experiments. I culled them down to a small group: five papers which I believed had greatest application to B2B marketing.[7–11] My reason for why these consumer goods studies can be relevant for B2B is simple: All human minds react similarly to stimuli. Remember the 2:00 AM alley cat story from Chapter One? Nearly all of us would have been emotionally alarmed in that situation. We can't help it. That's how our brains work. We're all human. And therefore, we share common responses to the world around us. In addition, several papers included in their experiments relatively complex, expensive, considered-purchase consumer goods: laptop computers, a sport utility vehicle, an Apple iPhone.

I analyzed the five studies with an eye toward identifying common threads of B2B suitability. Then, I summarized, condensed and re-condensed the learnings into conclusions, delivered (hopefully) with *brevity, clarity* and *impact* in this 3-30 message:

To maximize results with B2B products, use a combination appeal		
1. In specific situations, a **rational or emotional** appeal—one or the other—can be effective.	2. **Emotional appeals** to the senses grab attention, boost awareness and improve recall.	3. **Rational appeals—** especially for "thinking products"—move a prospect closer to a purchase.

That's it? Debate settled? Not exactly. But I consider the information to be of great value to B2B marketers. And for what it's worth, it's certainly different. I know of no case where information of this nature has ever been collected and analyzed for relevance to B2B, until now.

To help you put all of this into practice, we'll look at the discoveries written about in the papers. But first, a word about the main message above. It instructs B2B marketers to use a *combination* appeal, which, you should know, is my conclusion, synthesized from the data.

But how—*why*—did I come to this? None of the papers explicitly stated it. And only a couple even mentioned the idea that the two appeals could co-exist.

Here's how . . . and *why*.

The more I pondered the study findings, the more I became convinced of a simple fact. Most B2B marketing goals and situations seek the results delivered by rational *and* emotional appeals. B2B products, of course, benefit from attention, awareness and high recall. These results derive from emotional appeals. At the same time, fortunes improve for a B2B product when a prospect moves closer to a purchase—the results of rational appeals. B2B marketers want all these outcomes. And thus, my combination conclusion.

Research findings

Now that you understand what's behind my conclusion, here are the three Support Messages, along with their related Proof Points, all of this gleaned from the experiments:

In specific situations, a ***rational*** *or* ***emotional*** *appeal—one or the other—can be effective.*

Which appeal to use depends on the *product* . . .
A **rational** appeal may be more effective when the product is . . . • Highly differentiated from its competition • In the developmental stages of the product life cycle [8]
An **emotional** appeal may be more effective when the product is . . . • Not differentiated from the competition • In the mature stages of the product life cycle [8]
For **low involvement** purchases (products), emotional appeals have a more significantly positive effect on advertising attitudes.[11]
For **high involvement** products, emotional appeals are less influential.[11]
For **"value-expressive"** products (goods or services that express the personal values of the buyer/users) emotional appeals are effective.[8]

Which appeal to use depends on the *marketing objective* . . .
When the objective is to build awareness, knowledge and liking, **emotional** appeals will likely prove more effective than rational appeals.[10]
When the marketing objective is to move the brand into the consideration set or influence purchase decision, **rational** appeals will likely prove more effective than emotional appeals.[10]

Which appeal to use depends on the *audience* . . .
With "hard core" non-users, **emotional** appeals produced the greatest improvement in purchase intent.[9]
With younger consumers, **emotional** ads had a more significant effect.[10]
With older consumers **rational** appeals had a more significant effect.[10]
Emotional appeals had a greater effect on males than females. The authors noted this finding conflicts with commonly held beliefs.[10]
When the audience was not highly involved in the purchase or not highly knowledgeable about the product, an **emotional** appeal may be more effective.[8]

Emotional appeals *grab attention, boost awareness, and improve recall.*

***Emotional appeals* will likely prove more effective than rational appeals when the objective is to build awareness, knowledge and liking.[10]**
Recall of emotional ads was better than for rational ads.[7] • Consumers are prone to pay more attention to emotionally persuasive messages and pay less attention to rational appeals.[11]
Emotional ads, in general, outperform non-emotional ones in terms of generating ad appeal and initial brand image.[10]
Emotional ads were judged more original, unforgettable, entertaining, eye-catching and unique.[10]

Rational appeals—*especially for "thinking products"—move a prospect closer to a purchase.*

Audiences devoted more cognitive resources to relevant *rational* ads than relevant emotional ads.[7]
Rational ads are more effective at getting people to think more deeply about the product than emotional ads.[7]
Utilitarian (rational) appeals are effective when the product is utilitarian.[8]
Purchase interest was higher for product feature (rational) ads than end-benefit (emotional) ads.[9]
Consumers indicating intent to purchase the product, whether they did or did not use the product, reacted primarily to rational aspects of the ads.[9]
Informational (rational) ads rated significantly higher on ease of understanding, credibility, functional fit (appeal matches the brand) and informativeness.[10]
When consumers compare the advertised brand with other brands they know, they become increasingly rationally oriented.[10]
Rational appeals produced a greater ability for respondents to recite back more of the product's major advantages.[9]

Implications for B2B marketers

Unlike with consumer goods, those of us in B2B target, market, and sell products to people serving a role within a business. Whether it's the CEO, loading dock foreman, or an R&D scientist, rational and emotional thoughts enter their decision-making process. Emotional, because we're human. And rational because, well, it's business. And business deals with logical, rational things like goals, budgets, market share, quality, pricing policies, and the like. Keep this in mind as you develop your marketing appeals and creative approaches.

Final thoughts about rational vs. emotional appeals . . .

- **Which appeal? It requires judgment.**

  There are no absolute rules on when to use which appeal. The marketer's job is to navigate this conundrum, factoring in the marketing objective, along with what's known about the product, audience, competition and other situational conditions. This chapter, in fact the entire book, should help you sort through this.

- **Consider developing and testing several appeals.**

  Sometimes, you will know before developing your message and creative approach whether a rational, emotional or combination appeal makes the most sense. But sometimes you may not. In those cases, ask your agency or creative team to develop individual campaign concepts for all three appeals. Seeing the strategy carried out in differing creative approaches may tip the scales in favor of one of the appeals. If it doesn't, you can always conduct a concept test and get the feedback that will let you determine which appeal works best.

- **Practicality suggests a combination appeal.**

  Although the research papers I cited barely mention this, a combination appeal may be, in many or most cases, the most appropriate approach. In our many years of experience and practice, we've found that an emotional appeal can often be effectively carried out in the headline, main visual and major subheadings. Keep in mind, these will likely be the elements seen and read by the audience in the critical three seconds during which they decide to either stay with you or exit.

  In addition to grabbing attention, they should set the stage for the rational appeal that should follow and be carried out in copy, data and supporting information: diagrams, graphics, etc. Later in this chapter I'll show you real world examples of how our agency has done this for clients.

As we wrap up the debate, it's worth repeating: business, technical, and yes, personal issues enter into the B2B purchase equation—all the more reason to use a combination of emotional and rational appeals to emphasize your product's difference and relevance.

- **Enroll your people**

  On the people side of things, don't forget how vital your selling organization is to shaping perceptions in the marketplace. On-board your salespeople, dealers, distributors, wholesalers and key influencers to your product messaging and the support that backs it. Educate them on the rationale for the creative approach and appeal you've chosen for your marketing communications. Their support will multiply your efforts.

How to "do" creative

I began the chapter by setting the stage for the importance of effective and impactful creative work in B2B. You'll remember the most fundamental characteristic of marketing creativity—being novel, different or unusual in some way. Add in relevance and your creative work will deliver the complete package. But knowing the definition of creativity does not mean you know *how to do it.*

To explain how, I'm going to now show you how ***not*** to "do" it. I poke fun at a group of trite, undifferentiated ideas, tired approaches you've seen many times. These (un)creative idea starters are so generic, they could be used for just about any product. I give each generically bad idea a name, illustrate it with stock photos and provide trite, unoriginal headlines.

This isn't a database of every uncreative creative approach. (Not enough room!) Notice how the concepts do nothing to differentiate or convey relevance. My goal is to develop in you an early warning system that can root out bad creative work before it "***un***differentiates" your product.

After the "don't" lesson, I show examples of marketing creativity done the right way—work that differentiates, breaks through and delivers relevance.

We will close out this chapter with 14 things to know about putting creativity to work in your marketing. So let's get started . . .

When it's the same, it's not different. It's irrelevant, boring, repellant.

How many times have you seen these anything-but-creative ideas in B2B marketing?

Don't do this: The handshake . . .

— HEADLINES —

Trust. It's what sets us apart.

A commitment you can count on.

Looking for a partner? Then look no further.

Ahh yes, the old "put 'er there, pardner" idea. And did you notice how much irrelevant sameness you can create headline-wise for obtuse visuals?

Marketer, author, and renown copywriter Herschell Gordon Lewis, said, "A handshake is one of the most hackneyed, least inventive, least imaginative, and least creativity-indicative devices anyone can use as a main illustration."[12]

Commitment, trust, "partnership"—expressing these thoughts without corny words and pictures can be a chore. But it can be done. And keep in mind, when a partner invests money in an enterprise, they can make or lose money. When marketers use the word partner, they don't mean it in this way. But plenty of businesspeople do. When you use the word partner to mean anything but investor, you risk turning off business-owner types. Play it safe. Don't use the word partner to mean anything other than at-risk investor.

One final error to note for these words and images. They do not, *in any way,* identify the product being sold. As Herschell Gordon Lewis said in 2009, "Illustration should agree with what we're selling, not with the headline copy." [13]

Don't do this: The "Meeting" where our people provide solutions to "meet your needs."

— HEADLINES —

When it comes to solutions, our people make the difference.

Working together to meet your needs.

Looking for bright ideas? We'll bring teamwork to the table.

How many times have I heard this, "Chris, we like to think of ourselves as a solutions provider." Of course, you do. So does everybody else. "Solutions" depicted in marketing usually devolve into easy-to-dismiss mush. So how do you show yourself to be a consultative "solutions provider?"

A better question: why even try to show that idea? Nothing about providing "solutions" will differentiate you. Every going-concern business provides solutions to customers. Every single one. And before you rebut with "yes, but we provide *better* solutions," re-read Chapter One. Remember: Best or better do not differentiate.

It's okay to want customers and prospects to believe you're good at providing solutions. And it's okay to set that as a goal. But customers don't conclude that just because you say it. And that means it's *not* okay for you to say it.

A final thought: don't use "meeting your needs" or "meet your needs" in your copy. Vague, undifferentiating language like this wastes the time of your audience. And when you waste their time, they tune you out. Ask yourself, would one of your top salespeople say to a customer, "The reason you should do business with us is because we provide solutions to meet your needs." Top salespeople know better, good enough reason for marketers to avoid it as well.

Don't do this: The perfect presentation . . . where our people get down to business.

— HEADLINES —

Charting the future . . . together.

ROI—our promise to you.

In business today, you need facts from people you can trust.

Just look at those serious, "businessy-looking" actors with their serious, "businessy-looking" graphs. Somebody is seriously meeting somebody's needs!

Okay, enough with the sarcasm. Headlines? I tell copywriters, "If you need to say, 'Get it? Get it?' after someone reads your headline, you wrote a pointless cliché or pun." And that's the failure of the first headline.

Remember our discussion of direct and derived (generic) benefits in Chapter Two? Headline #2 goes straight to a derived benefit (Return on Investment), no mention of "how." At the same time, it does nothing to establish something different or relevant about the product.

And about that last awful headline: Hold off telling your audience what they need. It usually comes off as lecturing or arrogance—different, perhaps—but different not in a good way. Who wants to be harangued by a supplier? And finally, how about some bonus points for working an additional meaningless cliché into that last headline ("people you can trust")? Please don't say "trust us/me." People you cannot and should not trust say things like, "trust me."

Don't do this: Meet the experts . . . us!

— HEADLINES —

Solving your challenges with expertise.

A team of experts on call.

For solutions you can trust, talk to the experts you can trust.

Anybody ever call you an expert? Great. Do you call yourself an expert? Not so great. For years I've counseled clients: don't call yourself an expert. It's okay to highlight your expertise. Let others call you an expert. In our agency I made a rule for our people that states: "You don't get to call yourself an expert. If someone else calls you that, you may, with their permission repeat it. But under no circumstances do you get to praise yourself in that way. When you do that, people will consider you an arrogant ass. And no one wants to work with an arrogant ass." If you insist on going down this road, use the word "specialist," instead of "expert." This assumes, of course, you specialize in something valuable and relevant to your audience.

Headlines? "Solving your challenges" is an equally uninteresting way of saying "meeting your needs." Forget using both of those. And never forget buyers almost always buy what they want, and only sometimes buy what they need. This is important. I wrote a chapter about the difference between wants and needs in my first book, *Brand Busters*. What you learn there may surprise you.

The last headline expands on the "two-clichés-for-the-price-of-one" spree and gives you not one, not two, but *three* vague, uninspiring ideas in a single headline: solutions + experts + trust. Think of it as a lose, lose, lose creative approach.

Don't do this: The good doctor . . . with the perfect patient.

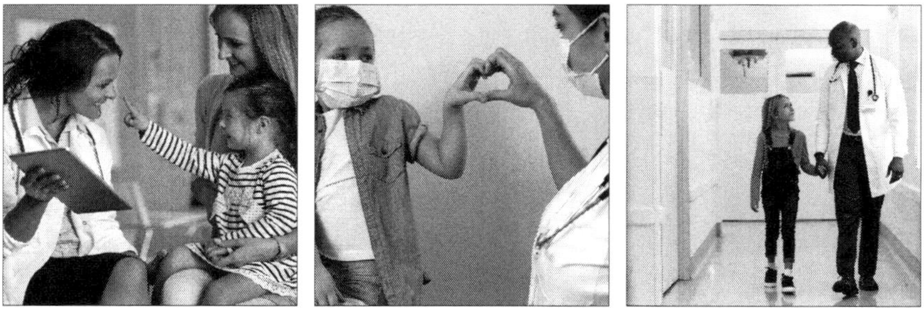

— HEADLINES —

Your patients and their families rely on you. You can rely on us.

Caring for patients takes plenty of heart.

We'll walk the extra mile with you.

Over the years, we've done lots of work in medical and orthopedic devices, pharmaceutical products, laboratory instruments, drug discovery and the like. As in other B2B fields, undifferentiating marketing flourishes here. And B2B marketers in other fields frequently depict a distribution customer with an end user, as in this case: doctor with patient. It's nothing out of the ordinary, as are the shopworn headlines.

These concepts try (but fail) to convey that the marketer really "gets it." They really, really, really, understand the audience and their world. But forgive me, if the marketer really understands all of this, why depict it in such a formulaic and simplistic way?

That last headline shows yet another cliché that begs a "Get it?"—sure sign of an irrelevant, undifferentiating idea.

Don't do this: The product as star.

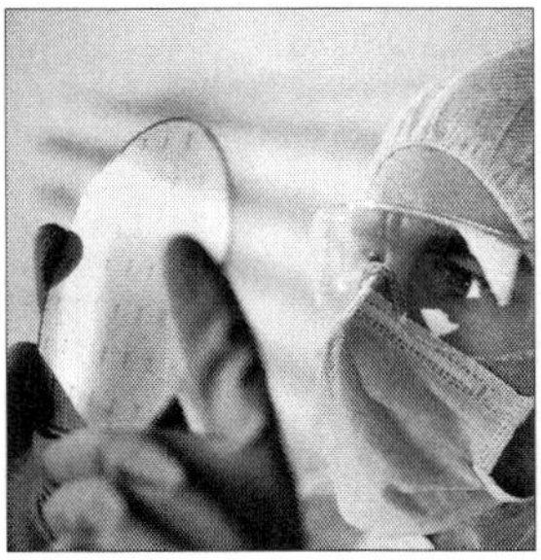

— HEADLINES —

Look closely and you'll see the difference.

Quality you can see.

Precision. It's built into every unit we make.

Nothing says "our gizmos meet your needs" like having actors point at or fawn over the gizmo. Just kidding. For an alternatively bad approach, just show the gadget and leave it up to the reader to figure out what makes it different and relevant. Concepts in this vein take a "here's what the gizmo *is*," approach instead of the more differentiating and meaningful "here's what the gizmo *does*" angle. These images drip with ordinariness, as do the mock(ing) headlines.

In the first headline, the writer appears to have no idea of what makes the product worthwhile. They leave it up to the reader. Sadly, for the marketer, no reader will hang around long enough to do that. In headline #2, quality takes center stage. Not the direct benefit of quality, mind you, just the word. Quality has been so overused it's all but devoid of meaning today. The last concept follows the "show the gadget" theme to its logical dead end. Buyers of machine tools and other B2B goods want the results of what a device can do. With this image and headline, it's anybody's guess what the contraption does for the customer. Yes, it's made with precision. But lots of machines are. So, who cares? Likely no one.

Don't do this: Tired analogies for service businesses

— CONCEPTS —

The Puzzle . . .

The Chess game . . .

The Symphony Orchestra . . .

By now, I bet you could have dreamed up headlines to drive these concepts even further into oblivion.

Banking, insurance, accounting, legal services, business consulting, IT, transportation and logistics—B2B marketers in these fields and more try to "sell" with analogies. But there's a big problem. Analogies confuse because they stray far afield from what the marketer actually sells. Show these images to 100 B2B buyers and ask them to guess the products being sold. They'll probably say: puzzles, chess games, and conductor batons, or musical instruments. That's how our brains work.

Participate with me in a mental exercise. Let's create a web display ad. We'll use the puzzle photo above as the main visual in an ad for a big, multi-national accounting firm. And for the headline, we'll use: "Solving the puzzle of multi-national tax strategy." Finally, let's stuff the ad with hundreds of additional words to explain the puzzle as an analogy for how the firm fits together all the complex pieces of international tax regulations into a profitable business strategy. Can you see the ad? To simulate how it will work in the real world, let's say we show it to 100 CFOs for three seconds. (Remember, people look at an ad for about three seconds before they either tune it out or continue reading.) Then, we'll ask CFOs what the ad is about. My guess is not a single CFO will say it's about multi-national tax advisory services. "If it looks like a duck and quacks like a duck, it's a duck." That's how minds work. Neuroscientist John Medina puts it more eloquently, "We learn and remember best through pictures, not through written or spoken words."[14]

When it's creative, it's different and relevant, and it captures attention.

To help you get the most out this, I selected real life examples for which I can provide background and context. As such, what follows comes from the creative folks at our agency, 5MetaCom. The creative work represents a range of products, situations and approaches for you to learn from. Ready? Let's take a look . . .

Do this, instead: The good doctor and patient . . . done differently.

Earlier, we saw phony-looking doctor-patient photos. Here's an approach that avoids that pitfall. Imagine you're a physician who must deliver bad

news to a young patient. This ad captures this emotion while also delivering information about a product that allows kids and their care givers to live a more normal life by testing blood at home. No more frequent trips to the doctor's office or lab. The combination emotional-rational appeal got noticed by many, including editors at *Business Marketing* magazine who said this about the creative concept:

> *". . .effectively uses emotion to communicate to its target audience of cardiologists the benefits of a home testing kit for their young patients whose blood must be frequently measured for its thickness. The photo of the little girl with the soulful, penetrating stare will draw doctors into the ad. An interesting photo of a human face has enormous stopping power."*[15]

Do this, instead: The product as star . . . done differently.

This creative example plays up the emotional effect of our client's product on their customer's end customer—the retail shopper. Our client designed and manufactured Point-of-Purchase displays for fast moving consumer goods makers like Procter & Gamble, Nestlé, and Pepsico. The cover image shows the benefit of their work: head-turning stopping-power. The rest of the marketing piece builds a rational case for how our client's displays deliver results in retail settings. As you might expect, it goes in depth to explain what's different about this display maker's methods and the relevance those methods have on ringing the cash register for consumer goods marketers.

Do this, instead: The problem . . . (and solution) as star.

Here's a headline and main visual we created for a promotional piece targeted to home inspectors, real estate agents, environmental consultants and industrial hygienists. It highlights a deal-killing, emotionally-charged obstacle that can derail a home sale: mold. In color, the "M" is made of green mold. The area above HOUSE shows the same green slime. It also delivers a rational solution: our client's product—a simple to use DNA test kit. With this device, real estate professionals can identify mold problems so they can be addressed before ruining a sale. The relevant problem-solution approach employs reason, accurate DNA testing, and emotion: a big hit to the pocketbook.

Do this, instead: Meet an expert . . . done differently.

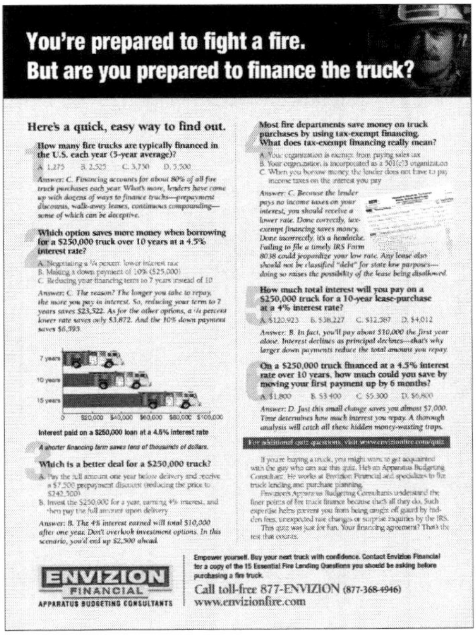

Rather than announce "we're the experts," this trade publication ad, targeted to Fire Chiefs, conveys this marketer's deep, financial knowledge. How? By educating readers on how to get more buying power out of their capital equipment budget. As one Fire Chief told me, "Buying firetrucks is about getting the most truck for the buck." And our client helped them do just that. In other words, they "sell" money—something often regarded as a purely undifferentiated commodity. As such, most Fire Chiefs shop for money on price. They look for the lowest interest rate. But our client, strategically, preferred to compete on value and expertise. And that's why we developed this rational appeal—a "test your knowledge" creative approach designed to make prospects think more deeply and look beyond interest rates when selecting a lender. The ad engaged readers to learn the direct benefit of our client's know-how: a better financial outcome and "more truck for the buck." And the audience responded. Readership scores tallied among the highest for any ad appearing in the publication. Do you have a special skill that helps customers get what they want? Don't say you're an expert. Instead, creatively convey your expertise. You'll help your customers and likely yourself.

Do this, instead: You may not trust us . . . so who do you trust?

This B2B marketer sells highly technical animal feeds to livestock producers. The creative concept we developed, however, doesn't highlight the product. Nor does it feature something commonplace like a customer testimonial. Instead, it showcases a veterinarian—an influencer who neither buys nor sells our client's product. So, what gives?

As most veterinarians will tell you, their clients (livestock farmers) frequently ask them for advice on animal nutrition. Producers see veterinarians as less biased than reps who sell feed. So, when a veterinarian vouches for a nutrition product, livestock producers pay attention. Strategically and creatively, the approach we developed interjects emotion—a warm photo, and a message of genuine thanks to these influential professionals. At the same time, it also delivers rational product information. To our knowledge, no nutrition company had ever shown a veterinarian in its advertising. This caused a stir in the industry, in a good way. That's what being different can do. And the approach achieved more. Using a real veterinarian, photographed in a real setting added to the relevance and credibility of the creative work in a way a stock photo and a "pseudo-testimonial" never could. To shoot this photo on location cost much more than a stock photo. But ask yourself, how much additional impact and believability did it add to the creative concept?

Do this, instead: A relevant, differentiating feature as star.

The cover of this literature piece puts the product's technical difference, and potency, front and center. Does it try to convince the customer this product is better or best? No. This product competed in a segment crowded with rivals, competing opinions, chatter and debate about which one was "best." Our creative concept allowed our client to avoid that trap and compete from a differentiated position of strength. Among "me-too" entrants, only one product was, in fact, the most potent and powerful—a rational and emotional appeal. Thanks to that laser-focused concept, the audience can spot the relevant difference in the product. Potency became a standout differentiator, one that piqued the audience's curiosity and set the stage for delivery of the product's compelling technical story in the pages that followed.

Do this, instead: Meeting your needs . . . with utility.

Instead of generically saying "we meet your needs" why not dramatically show what makes your product different? Show how it solves problems in a unique and relevant way. That's exactly the intent of this in-store poster we created. Background: our client's seed came to the farm equipped with a unique feature—a protectant coating. This safeguard enabled seeds planted in early spring to better survive and thrive in cold, harsh, wet conditions. No farmer would purposely plant in those conditions due to fear those seeds would never survive. But showing how they *could* do this dramatized the product's difference. It gave farmers the reason to believe. And this shocking, yet relevant visual—a planter in the snow—hammered home the seed's unique utility.

Do this, instead: Can the name say it all? Maybe not, but...

. . . it can say plenty. When you're a budget-tight startup, competing with global giants, getting people to understand what you sell, where to buy it, and why you're different can be a monumental challenge. Here's an example of a brand name and logo we created designed to achieve all three. Our client, iCorn.com, sold corn seed direct to farmers on the internet. And they were the first—a big difference. In our strategic planning with the client, we all concluded the brand name and logo needed to do extra duty in the marketing mix. The creative solution—name and logo—not only tell you what they sell, they telegraph the company's difference: you can buy from them on the web. Finally, name and logo combine to deliver the web address which tells people where to buy product. The right name and logo can be powerful marketing tools that differentiate and deliver key information. What could be more efficient?

Do this, instead: Present facts; let the audience draw conclusions.

Here's a page from a sales brochure we produced to launch the new category we discussed in Chapter Four: waterless fire protection. The image shows an emotional scene, a drenched and destroyed computer room. And what destroyed it? The very tool put in place to protect those assets from fire: water. The copy, in this interior page of a collateral piece, underscores differences between water sprinkler fire-fighting and our client's waterless technology. As the headlines says, "Water can destroy critical business systems and irreplaceable assets." The body copy presents rational facts about sprinkler systems and water:

- Water is not released until a fire is large enough to do substantial damage.

- Water is electrically conductive so it may damage sensitive electronics untouched by fire.

The emotionally-charged photo sets the stage for presentation of rational facts that allow the reader to reach a conclusion about which approach, water or waterless, provides the utility they're after. Do we present every difference between the two? No. As marketers we select which facts to include. That selection process is part strategic and part creative and those choices will and should be included in a product's Marketing Message.

Creativity in your marketing communications: 14 things to know . . .

1. Creative work that strays from the strategy

The biggest mistake we see is—creative work that strays from the strategy. And it's easy to understand why this happens. We commonly ask creative teams to diverge, cut loose, let their minds wander, think freely. This approach usually produces a wide range of concepts, from creatively and strategically brilliant, to way off-base. But you know what? That's okay. It's part of the process. Off-base ideas bubble up in spite of how much direction we provide. And be warned, if you make a habit of scolding a creative team every time they miss the mark strategically, you run the risk of shutting off their idea fountain. We all want amazing, differentiating, super-creative ideas. And, of course, we don't want creative work that misses the mark. Yet experience tells me great creative work comes packaged in with a bundle of not-so-amazing ideas. Remember . . . that's okay! But, I didn't always understand this.

I used to drown creative teams in loads of background information and detailed instructions. I thought the more explanation I provided, the less likely they would ever produce an off-base idea. But it hardly ever worked that way. In time, I learned a better way. Provide the team with a creative brief—the key facts—along with a solid 3-30-3-30 Marketing Message. The conciseness does a way better job of jump-starting the creative process. Google "creative brief" and you'll find plenty of good examples you can use or adapt. Next issue: How should you evaluate the creative work? How will you know when an idea hits or misses the mark? Simple. Refer back to the Marketing Message. Use it as a both a target and measuring stick.

Recently, I took part in a concept review session with a creative team. In our shop, we refer to a critique session of initial rough ideas as an "opening argument (OA)." In this OA, I saw some really creative stuff. Most of the rough concepts presented did a good job making the case for our client's product over their main competitor. Great work, but off track in one important way. The marketing strategy and message weren't intended to convince the audience of our product superiority. Rather, we wanted people to take away the message of "most preferred and fastest growing." Going back to the Marketing Message in our review process spurred some terrific second-draft work on the project. Often, as in this case, it takes iterations and refinements to produce creative concepts that hit the mark creatively and strategically.

2. Money follows ideas

"There's no shortage of money, only a shortage of ideas." Have you ever heard this? I didn't expect you would. It's one of my favorites. I'm not sure if I made it up or borrowed it. No matter. The point is, don't get dragged down into thinking your boss, "the company" or your client won't go for a big idea because of cost. Am I saying to routinely disregard budgets? No. You should always try to "do" great creative work that stays within the budget. But sometimes a market-shaping idea also warrants thinking outside the budget. Am I saying groundbreaking, budget-busting ideas will always get funded? Not at all; most times they won't. But every once in a while, they will. Don't let fear of money shut you down. And don't let fear of "they'll never go for it" prevent you from pursuing big, truly differentiating ideas.

3. Pick a goal

What do you want the reader or viewer *to do*? It could be to remember something (such as your 3-30 message). Or visit a website . . . look at a video . . . download a white paper . . . enter data into an economic calculator, etc. The point is that you should know what action you want the audience to take, and make that part of the instruction you provide to the creative team. Don't make them guess about your objective. Be explicit.

4. Use your Marketing Message—Don't *change* it.

And don't simply regurgitate it. *Do* dramatize it, illuminate it, express it in a captivating way. To explain what I mean, I'll use the ad I showed earlier, the one with the young girl. The strategy direction and main Marketing Message (to physicians) about this product went something like this: "Your young patients can now have their blood tested at home." An uninspired, non-creative concept might simply have restated the main message in a headline such as this: "PT testing your patients can now do at home." For a visual, the art director could have shown a large photo of the blood test machine. Close your eyes and you can probably picture this ad. And in B2B marketing, you've no doubt seen a similar "show the gizmo" approach repeated *ad infinitum* in scores of product categories. Does the bland ad convey the message? Yes. Is it strategically correct? Yes. Does it grab the audience and draw them in? Provoke further thought? Not in my opinion. Powerful creative concepts do more than reiterate. They breathe life and differentiation, emotion and powerful logic into a Marketing Message.

5. Do your Homework

There are known, proven processes for developing ideas. Learn them. Follow them. Practice them. Writing about ideas in *Systematic Approach to Advertising Creativity,* author Stephen Baker commented, "More often than not, they are the end product of arduous and well-organized intellectual activity." Continuing, he said:

> "Information gathering comes first; analysis follows. And finally—eureka!—the idea 'occurs.'"
>
> "Anyone who can talk is able to write. Anyone who can see is able to visualize. *And anyone who can think is able to have ideas.*" (my emphasis)
>
> "Once you are willing to concede that creativity is not a haphazard but a purposeful exercise, things fall into line with surprising ease."[16]

6. Good things take (a little) time.

Even as little as a couple of days between receiving a project assignment and starting on the task allows the mind to subconsciously kickstart the creative process. In my own creative work (writing, strategy development, business planning) I find even a day or two of inaction between project kickoff and commencing work dramatically increases the quantity of ideas and quality of the output. Give yourself, your team or your agency at least a little fermentation time for ideas. It will pay off.

7. Critique things early

The best creative direction in the world won't contain all possible information a creative team could ever need. And so, the first rough concepts you see likely won't address everything they should. I know this because I once invented what I thought would be a process for writing the perfect creative brief. In my logic, a perfect set of creative instructions would yield "perfect" creative work. Wrong. I even hired a PhD with science credentials to write the danged things. But no matter how much "good" information and direction we provided, we unwittingly left stuff out. Invariably, creative people conceived of things we never could have imagined. And in doing so, they strayed into areas for which we had not provided information or direction. This always came home to us when we saw first draft creative concepts. After more than a year trying, I abandoned the quest for perfect instructions in favor of something simpler and more effective. It works like this:

One: Give the creative team a summary of key information you think they'll need. Your best guess is all we're after.

Two: Ask the creative team to work on the assignment for a short while and develop rough concepts.

Three: Convene a meeting of the creative team and reviewers. We have rules that specify who reviews initial creative work.

Four: Review rough concepts. Discuss them. Talk about what's good, not so good, what's missing, what should be added, subtracted, etc. At this point, you'll almost always discover the creative team needs additional information. That's okay. This is the point at which you provide it.

8. One main idea for your concept!

A piece of creative marketing communication should center on one main idea. That's the beauty of using a 3-30 marketing message to drive the creative work. It has a main message: the :03 Headline. You can't make dozens of points and hope to break through to an audience. Here's an analogy. You can lay on a bed of nails, right? None of them poke through. But try laying down on a single nail. It breaks through (and bloody-well breaks the skin). Same concept goes for creative work. A single, sharp point stands a better chance of breaking through. Creative concepts muddled with multiple ideas and messages cannot, and will not, make an impression. Neuroscience data I explained earlier showed that you only have 2.75 seconds to grab someone's attention, slow them down and get them to recognize there's something different and relevant going on. To poke though, use a single nail!

9. A word about pictures . . .

It bears repeating, "We learn and remember best through pictures, not through written or spoken word," says my favorite neuroscientist, John Medina. Here's more from him:

> "If information is presented orally, people remember about 10 percent, tested 72 hours after exposure. That figure goes up to 65 percent if you add a picture."
>
> "Put simply, the more visual the input becomes, the more likely it is to be recognized—and recalled."[17]

Harness the knowledge of how our brains function in your creative work. The "real world" examples discussed earlier show visuals hard at work conveying messages more quickly and powerfully than words ever could. Even the fire truck ad, with its hundreds of words of copy, uses visuals (a graph) to convey key messages. Remember from history the cave drawings of early humans? Pictures, visuals and symbols preceded language in our species' evolution. That visual wiring exists in the brains of your prospects and customers. Leverage it.

10. About stock photos and design templates

Stock images and design templates have made the creative team's production job much easier and less costly. But with this good, comes bad. These plentiful tools have also made the marketer's main job—differentiation—harder. After all, every creative team has access to the same image and template libraries. This means looking professional, image-wise and design-wise, has never been easier. It also means differentiating—looking, sounding and feeling different to your customer—has become harder. You and I can spot "made from scratch" vs. "generic" a mile away. So can your customers.

By the time you get to the creative work, your differentiation strategy and message should set you on a differentiated path. When that's the case, well-chosen stock images, will not impair your ability to stand out. My advice, however, is to go easy on stock creative assets for your biggest, most differentiating ideas and marketing materials.

Launching a new product? A new website? A new campaign? Give it a differentiated strategy and message, a unique creative approach and a custom, one-of-a-kind feel with commissioned images and creative work.

Will it take more time? Yes. Cost more money? Of course. Will the payoff be worth it? That's for you to think through.

The point is to do just that: think it through. Don't automatically reach for the quick and economical stock creative solution. Armed with a solid, differentiating strategy and message it's perfectly okay to use stock creative assets. But don't do what you saw in the examples earlier in this chapter—non-existent strategy plus generic messaging coupled with unimaginative stock images. In all cases, search for images and designs that make you look different and relevant. Those stock assets are out there. Don't just settle for "put 'er there pardner"-handshake type images. Nothing could make you look less different and special.

11. Verisimilitude: For it to *be* true it must *look* and *feel* true.

The great marketer and copywriter of Herschell Gordon Lewis (look him up; read his books), taught the concept of verisimilitude. The big word simply means, "the appearance of being true or real." I once had the good fortune of attending a copywriting workshop session with Herschell. From that day on, his concept of verisimilitude has stuck with me.

From a practical viewpoint, the principle means everything in your marketing must be congruent. I'm talking about all components in your marketing mix: the product, its name, messages, packaging, pricing, promotional approach, giveaways and promo item, trucks, buildings, uniforms, collateral materials, your website, etc. Everything must look and feel the part. For example, let's say you wish to differentiate yourself by being highly knowledgeable on some key technical issues relevant to your customers. If so, your people should be well trained on the topics. Your website should have plenty of content about them—white papers, interviews, diagrams, calculators, etc. Yet, I've seen marketers say, "We're the experts on this topic," and then display the complete incongruity of handing out a one-page flyer on the subject. And when you go to their website? There's not a shred more information about it. A one-page flyer!?

Give me a break. Everyone knows a true and real authority has much more to say than that. Customers see right through this kind of phony marketing spin. So how do you use verisimilitude in your marketing? Time for examples:

Example One: To build credibility

One client of ours produced annually a 100+ page data-dense compendium of performance results from tests they conducted on their products across a wide range of applications. Tests were performed anew each year. And each year a new compendium of technical data was compiled, published and distributed to sales teams, customers and prospects.

It was an arduous and expensive endeavor. And its contribution to sales was constantly debated. Some customers appreciated the information; they used it to make purchase decisions. Others viewed it skeptically, with comments such as, "This is your data, your tests . . . of course your products show positive results." True enough. The data came from applications in which the products performed well.

Each year in budgeting, we wrestled with whether to continue or eliminate the project. One year, our agency hatched a third option. Instead of "go/no go" what if we tried to improve the selling power of the data? Enter the principle of verisimilitude and a creative concept that totally reimagined the dry technical publication into a magazine, complete with a name and masthead, dramatic color photography, feature articles highlighting R&D projects and people, interviews with customers, and reformatted, easier-to-read and easier-to-understand data tables.

We intended to raise the believability of the data, and by doing so, get more customers to use it when making purchase decisions. Based on the creative idea, our client took the leap of faith. What's more, they piled on with an additional big idea. They decided to include data on applications where their products did not perform well. They hypothesized that data showing poor performance would boost perceived legitimacy of data showing good performance results and thereby boost sales in those applications. Did it? There's more.

One day, in the throes of creative development for the magazine, an idea, spurred by the verisimilitude principle, hit me. Staring at a cover mockup, I pulled out a marker and scrawled, just under the masthead and opposite the publication date, this: $5.00. After all, we wanted to enhance the perceived value of the data. We wanted the magazine to appear more valuable than an ordinary piece of marketing literature. So, I put a price on it: five bucks. A revised mockup with a $5.00 price on the cover produced a smile from my client. Game on.

We rushed to get the magazine delivered ahead of our client's largest trade show. To enhance verisimilitude, we purchased magazine racks and filled them with the newly printed magazines. I attended opening day, intent on learning if the magazine would have any effect on customers and prospects.

To my amazement, just a few moments after the show floor opened, a prospect sauntered in and made a beeline to one of the racks. Approaching, he gazed intently at the cover. He picked one up. He thumbed through. He paused. He looked closely at several pages in succession. Abruptly, he stopped. He tucked the magazine under his arm. He reached around to his back pocket. He pulled out his wallet. With one motion, he opened it and picked out a bill—a $5.00 bill. He walked directly to the counter (manned by my client, Steve) held out the $5.00 bill and said, "I'll take one of your magazines." Before Steve could react, the man spun on his heels and walked out of the booth with his nose still stuck in the pages.

I wish I could say everyone who walked into the booth that day forked over five bucks; they didn't. But some did. And to each one, our client's people interjected, "It's yours. Today, they're on the house." The payoff of our effort to build credibility for our client's technical data had begun. Over time, buyers in this industry relied increasingly on information our client continued to publish each year.

And we witnessed firsthand the power of verisimilitude to change perceptions.

Example Two: To bolster the brand

I've explained how our agency helps clients differentiate technical and scientific products. That's what we do and what we have done for 40+ years. And yet we also know it's important to look and behave congruently with that brand position. Take a look at the photos below. You'll find these signs mounted on the wall outside several conference rooms in our offices. Notice the names, the element numbers and atomic weights? The type of sleek materials from which they're made? Chalk all of that up to verisimilitude.

We work with marketers of technical and scientific products. We provide a professional service—an intangible. Yet tangibles surround our brand—things like our name, office décor, website content, the type of knowledge and information we possess and pass along to our clients, including intellectual properties, such as this book. It must all fit together and co-exist congruently with who we are as an organization and what we provide to our clients. I remember one PhD chemist who visited our offices commenting to a colleague while staring at the Chromium sign, "Atomic weights. Maybe these guys know their stuff." Our conference room names and signs are just one element of verisimilitude employed to communicate our brand.

Conference room signs at the offices of 5MetaCom. How well do they convey the company's brand?

12. Make your website mainly about content and organization.

A website succeeds because it provides value to the user. And value in a website usually comes down to content and organization. Good content and good organization delivers good value, and vice versa. Lots of folks talk about "website design" and of course, design is important. But content and organization are where the action is when it comes to B2B marketing. I can't tell you how many times I've talked to folks who say, "We do website design." Then when I ask how they develop the site's organization, content and copy, I mostly hear, "Well . . . we don't actually do that. That stuff comes from our client." And what is it you actually do?

If you work with someone like that, understand they're a decorator, sort of like a home painter who takes your existing home (content) and pretties it up. Don't get me wrong. A nicely painted house is a good thing, but it can't make up for a lousy floor plan, badly sized rooms, lack of architectural pizzazz, etc. Stop worrying so much about the prettiness of your website. The layout, colors, typefaces, photos and illustrations are there to support and enhance strategy, organization, content and copy. And what should the content and organization do for you? Express your difference and relevance to site users! Your site should be a key component in your strategy of differentiation and relevance.

13. Learn the science and craft of marketing communication, creativity and persuasion.

Lots of smart people have researched how to improve the effectiveness of your words, pictures and ideas. Google them. Study their work! Here are a few to get you started:

When it comes to **readability and comprehension** of your written words, turn to Rudolph Flesh.

For the art of **persuasion**: Dr. Robert Cialdini.

For how the **brain functions** and reacts to information (including marketing communications): Dr. John Medina.

For **creativity** and the process for developing creative ideas: James Webb Young and Dr. Edward de Bono.

About the field of **behavioral economics** (why and how people make decisions and spend money): Dr. Daniel Kahneman.

I could go on and on. But these will get you started. Understand, these people are by no means the only authorities with good ideas. Hundreds of people write, speak and publish stimulating ideas about marketing. Be curious. Become a life-long learner of your craft.

14. The importance of names

I could write an entire book on names and naming. Properly approached, names can be one of the most important and effective tools in your differentiation arsenal. And I'm not just talking about product brand names. Recall the chapter on categories? Yes, categories require names. And as you saw, an effective category name will contribute greatly to your marketing success. What else can you or perhaps should you name to hammer home your marketing differentiation? Here are just a few:

- **The company**
- **A division or unit**
- **A team within your business**
  Example: "Methods Improvement Team," which I'll explain in the next chapter.
- **Products/services**
- **A physical space or building**
  For instance, to make a point about differentiation, should you name your R&D Center? We did this once for a client.
- **Categories**
  We once invented a name for a new field of science that highlighted our client's unique technical capabilities.
- **Processes**
- **Problems, diseases, issues**
  (the pharmaceutical industry does this all the time)

The possibilities for what you can name to differentiate are practically limitless. But don't go overboard. Use them sparingly, for emphasis. When you name inconsequential things, you'll likely diminish the power of your named differentiators and risk looking phony. As someone once said, "When you emphasize everything, you emphasize nothing." The best names hold up technically, scientifically and marketing-wise.

And how important are names? Consider this. I'll wager you're familiar with something known scientifically as a "gravitationally completely collapsed object." That's the original name used to describe what we all refer to today as a "Black Hole." The astronomical concept of a black hole had its origin in Einstein's General Theory of Relativity from 1915. And there, in complex theory, it languished, at least in the eyes of the general public. That is, until 1967, when theoretical physicist John Wheeler gave it a new name: "black hole." The short, concise, interesting descriptor attracted attention and quickly gained popularity, to the point where today hardly any person remains unfamiliar with the term. That's the power of a good and differentiating name.

CHAPTER 7

15 Differentiation Idea Starters —and 6 to avoid

When compared to consumer goods, B2B products can be difficult to differentiate. Consumer product makers can sometimes make a simple change such as packaging, a new flavor or scent, or new colors to quickly launch a new, differentiating idea into the marketplace. With B2B products, especially those in highly regulated industries, even a slight differentiating change to an existing, approved product can be prohibitively expensive and time consuming. With that in mind, most of the idea starters below require no changes to your product.

The key to making any of these ideas truly differentiating is to back them with hard proof. Differentiation in B2B isn't about marketing spin. It's about facts, utility and relevance. For instance, let's say no other products works the way yours does. If that's the case, you need to do more than simply say those words. You must credibly support that claim and clearly explain it technically.

Now that we know the ground rules, consider and explore the following concepts when developing your point of differentiation. Use any or all of these to kickstart ideas:

1. It's self-evident: New! is always different.

When something is new, it's no longer like it was. That makes it different. So, when you have something new—something relevant to the customer—don't be afraid to say it. The new thing could be about your company, your product or just about anything else. Of course, as with all things B2B, new also can mean untested, unproven, and risky. Address that hurdle. Provide credible, risk-mitigating facts. It could be detail about how you make the product, testimonials from influential users, test results, analytical information or other facts that reduce perceived risk and allow for your "new" to be seen as positive and differentiating.

2. How the product works

My guess is your product doesn't work *exactly* like every other one you compete with. If so, use this to your advantage. Show how your product works in video, animations, via test results, or in the words of a satisfied user. You can explain how your product does what it does in countless creative and interesting ways. Recall the example from Chapter Two that cited market research which uncovered how a product worked as a market-moving differentiator? That's the idea here.

3. How it's made

I see this as a great untapped area of marketing firepower. Lots of B2B marketers could and probably should use it to differentiate. Occasionally I see marketers make a stab at this. But they often miss the mark. They fail to translate a different way of manufacturing (a feature) into something relevant to the buyer. Not too long ago, in a discussion about differentiation, a very smart client of ours offered up a fill-in-the-blank statement he learned as a young B2B sales rep that taught him how to do this. The phrase he uses to this day goes like this:

Because of: ________________________________,
feature

you will be able to: ___________________________,
benefit

which means: _____________________________.
relevant outcome

4. Problem-solving ability

Once, I took a trip through the plant of a metal fabricator client with Paul, the company's sales and marketing director. Mid-tour, he turned to me and said, "Detour . . . this way." He darted through a wide doorway under a large sign that said, "Engineering Department." Inside, Paul sidled up to a CAD workstation and asked the engineer manning it a question, something about a big customer project they were bidding. After a couple of back and forths and a few nods, the two landed on a resolution. Just as quickly, Paul pivoted back into the hallway and continued the tour. As we walked, Paul explained how Kevin, the engineer, had come up with an idea for a slight redesign to the customer's blueprint. The tweak added strength to the part and also reduced manufacturing cost. "Is that a special service you provide just to certain customers?" I asked. "Nope," Paul said. "We do it for any customer. Look, when we have an idea we think will improve a part design or reduce cost, we bring it up." By "all the time" he really meant it. In their world, customer blueprints first went for a "problem solving" engineering review before moving to the estimating team. "Do your competitors do this?" I asked. Paul said he didn't really know. And that was that, for the time being.

Later, as our agency worked on a new marketing campaign for Paul's company, the scene I just described came flooding back. How could we make the "service" I'd witnessed into a relevant differentiator? Could we somehow highlight their problem-solving capability and market it? Could we brand it? Give it a name? Highlight the engineering acumen they delivered to their customers? Yes, yes, yes and yes. And that's exactly what we did.

We coined a name for the engineering group and their ability to enhance part designs. We called it the "Methods Improvement Team" or "MIT." (Yes, the "engineering" flavor of the acronym was intentional.) You'll remember Paul said he didn't know if his competitors offered something similar. In the end, this ended up not mattering at all. No competitors talked about it in their marketing. Our new campaign did. It went into detail about the Methods Improvement Team, its relevance and benefit to customers. And our client became known as the go-to problem solvers in their segment. And rather than compete on price, our client's sales and technical people became engaged in discussions about improving manufacturing efficiency and performance of their client's parts, a positive direct benefit for both the buyer and seller.

5. Non-manufacturing process(es)

We used a "process" differentiator for a client that made components and sold them to U.S.-based original equipment manufacturers (OEMs) of heavy equipment. Our client faced competition mainly from competitors located outside the United States. It gave those challengers a price advantage that had begun to put pressure on our client's margins. As part of a brand differentiation workshop, our agency conducted discovery research interviews with their customers. One key finding stood out. Many customers remarked, unaided, on how easy it was to do business with our client. They told us how the company answered phones promptly. They explained how much they liked being able to talk with knowledgeable employees, instead of "call center" contractors. Most described these people as helpful and well-trained. Many remarked about the lack of language barriers, a big difference compared with several other offshore suppliers in the category. Nearly every customer said this personal service approach set the company apart. One customer told me he liked it that real live people actually answered the phone. No automated answering and no drawn-out series of keypad punching to get to a real person. And no obligatory "please listen carefully as our menu options have changed" warning. Almost to a person, customers voiced appreciation for how quickly and efficiently our client's people provided answers.

Our interviews also covered products. In this area, however, most of the products—our client's and their competitors'—were seen as similar. The surprising thing about what we learned was that our client had been making no mention of their highly differentiated service—real, live people, knowledgeable and helpful people—not on their website, nor in any of their marketing materials. The marketing message we developed addressed this. Yet if not for the willingness of the client to pursue new ways of communicating their service difference, their sameness would have continued. Even more scary was something the company CEO told us. He said they had been leaning toward discontinuing their employee phone support and farming that work out to an off-shore call center as a cost cutting measure. "Without the differentiation research we would have made a grave mistake," he explained to me later.

6. What others say: authority and consensus

How many times have we all seen marketing copy like this?

- Used in more Fortune 500 companies than any other . . .
- Ranked #1 by JD Power . . .
- 5,263 five-star reviews and counting . . .

And why do we see so much of this? Because it influences. It persuades. It works. Don't just take my word for it. This fact comes straight from PhD psychologist, professor, author and world authority on persuasion and influence, Dr. Robert Cialdini. If you haven't already read his international best-selling book *Influence: The Psychology of Persuasion,* you should. The book has sold more than 5,000,000 copies worldwide. See what I did there? I used authority and consensus to build a case for Dr. Cialdini. Every marketer should learn how to do this.

Speaking in 2019 to a group in Seattle, Washington, Dr. Cialdini said this about authority and consensus:

> "When people are uncertain, they don't look inside themselves. They look outside for answers. One of the places they look is authorities . . . the other place they look is to their peers, people just like them. People follow the lead of *many others* and *similar others.*"[1]

Market/leadership, #1, best-selling, most-used, most preferred, chosen by ________ (authority) as ________ (something relevant to the customer).

These are all valid ways of expressing authority and consensus. Yet marketers frequently fail to harness this power. I gave you a taste of how to do this above when talking about Dr. Cialdini. For another example, I'll use my earlier book *Brand Busters*. If I ignored the concepts of consensus and authority, I might say something like this: "*Brand Busters* is a great book; every B2B marketer should read it." But I know that's not how to differentiate and persuade. So, what if instead, I said:

Brand Busters: Did you know?

- Chapter 4, on market segmentation, was awarded a University Classroom Resource designation by the Advertising Educational Foundation (AEF).
  - ✓ The AEF is the educational foundation of the Association of National Advertisers, the U.S. advertising industry's oldest and largest trade association.
- As of this writing, only 100 book chapters have ever received the Classroom Resource designation since the AEF's founding in 1983.
- After launch in the U.S., *Brand Busters,* was published in India by MacMillan Publishing:
  - ✓ MacMillan is one of the top 10 largest global book publishers.
  - ✓ India is the second largest English-speaking country in the world.

Do the preceding bullets make a more meaningful, relevant and persuasive case than simply saying *Brand Busters* is a great book? I'll let you decide. And did you notice I used a "Did you know?" technique to draw you in? How you creatively introduce and present your case is also a part of the persuasion equation. I aimed straight at your innate curiosity as a reader of marketing books. There are numerous other ways to create interest and engage your audience.

For buyers, B2B purchases involve managing risk and uncertainty. And that's why including what others say is usually a good idea. Does a recognized industry authority advocate for your product, your approach, or your point of view? Include that in your marketing. Do lots of your target prospects' peers use your product? Then say so in your marketing.

7. The inventor (the first)

As I mentioned in Chapter Four, Mercedes-Benz once ran a TV ad campaign that said, "We invented the automobile." I found this to be a fascinating and powerful way to convey know-how and leadership. Years later, it has stuck with me. Has your company invented something that caught on? Talk about it. One client of ours invented a technology in the early 1900s that revolutionized how materials were strengthened and enhanced. Their innovation paved the way for things like high-performance jet engines. Do we talk about that in their marketing? You bet we do. And don't forget the additional ideas discussed in Chapter Four about how to use "first" as a marketing differentiator.

8. Lots of experience

Perhaps you have more experience than competitors. Done properly, this can be a differentiator. Unfortunately, many marketers communicate this idea neither properly nor to their advantage. I often see marketers state experience in a way that simply says, "We're better than the other guys." The right way? Here's how we talked about experience for a client who supplied products into the aerospace market: "At any given moment, more than a million pounds of our materials are aloft in the skies somewhere in the world."

9. One-of-a-kind perspective

Take a look at this copy I wrote for a life sciences start-up, a company of novel thinkers with zero experience:

> *"**Old Way—the insiders look outside. New Way—the outsiders look within.** Intent on applying new science and new thinking to the field, these industry outsiders pursued a novel approach to the science . . ."*

The rest of the copy explained their unique "outsider" point-of-view, a perspective that freed them to dream new possibilities in a way long-tenured suppliers couldn't conceive of. Rather than trying to "fake it," the copy cheerily confessed to their lack of experience. With credibility as truth-tellers established, the rest of the story went on the explain how these rebels invented a new field of science. Being an outsider might have conjured up concerns among buyers. At the same time, it also conferred on them a unique "outside the box" point of view that competitors could never claim. This differentiated our client's people, their science, and their products.

10. Add something intangible

What about a new technical support service, an extended warranty, a fixed trade-in value, or an improved user's guide that helps customers get greater utility out of the product? The possibilities for adding an intangible are endless. And when you do add something, make sure buyers know about it. For a client that added a performance guarantee, we recommended a change in the brand name. Physically, the product remained the same. The new performance guarantee, however, meant customers received greater value. Using the old name would have hidden this. So, we developed a new one. We also created a new logo and included in the design a distinctive visual cue to trigger recognition of the differentiating guarantee. Ask yourself: What can we add? And: How will we draw attention to it?

11. Shape, size, dimension, weight, etc.

Where I live, White Castle sells square hamburgers in square buns. Similar bread and meat to other hamburger sellers, just a different shape. Is yours the only oval-shaped widget? We once highlighted the unique, curved shape of one client's product. Is your product more concentrated (so you add less, buy less, store less)? Or perhaps less concentrated (so it's easier to measure and mix properly and therefore produces a more consistent end product)? What about color? To highlight a client's product reformulation, we recommended they change the hue of the product colorant. Regulations required color be added to distinguish the product from water. Regulations did not specify any particular color, and we wanted people to know the product was new and different. "Look for the new, blue formulation," became a standard sign-off in our marketing. You don't need to change size, shape, dimension or color. But when there's something inherently different about the physical qualities of your product compared to others, explore expressing it as a relevant and meaningful differentiator.

12. Highlight a negative; include both sides of an argument.

A television campaign for a mouthwash once touted its bad taste. It cited this negative as proof for how powerfully it cleaned your mouth. "Tastes so bad it's got to be good," went the tagline. Ad researchers call citing both positive and negative attributes a "two-sided" argument. A one-sided argument only mentions positive attributes. One research study, using a consumer product, compared the effects of one-sided versus two-sided advertising. Results showed:

- Two-sided (vs. one-sided) advertisements increased perceived credibility of the advertiser.
- Mentioning negative product attributes created more favorable attitudes toward the positive attributes mentioned in the ad.[2]

For B2B products, credibility of the seller almost always enters into a purchase decision. So, when you want a credibility boost for your company or product, consider incorporating two-sided arguments. Admitting negatives produces the additional benefit of increasing the perceived truth of positive claims you make. Let's say your major point of differentiation seems far-fetched or implausible to your audience. This would be the perfect time to employ a two-sided argument in your marketing. Most B2B marketers shy away from highlighting a negative. Chances are, a two-sided argument will make you look much different than your competitors.

13. Cite unusual or high profile uses or applications

One of our client's building materials was selected for installation in the One World Trade Center building (which replaced the fallen World Trade Center towers in downtown Manhattan). This product was the only one in its category selected. You can bet we made that the centerpiece for a differentiating ad campaign that featured both emotional and rational creative appeals.

14. Verisimilitude

Earlier I went into depth about "the appearance of being true or real." I mention it again, so you won't forget it. Remember the poster with farm tractor and planter in the snow from Chapter Six? As an extra measure of differentiation, we piggybacked off that point of purchase concept by creating a snowball toss game for our client's trade show booth. Full disclosure: we found out real snowballs didn't hold up well at their outdoor mid-summer trade shows, so we ended up substituting baseball-sized ones made of white Styrofoam. Despite the initial foul ball, turns out people had fun with the lighthearted kitsch. And the tie-in to snow added an extra measure of energy behind our client's product differentiation in a crowded trade show environment.

15. Specialization

All other things being equal, specialization connotes expertise. Ask yourself, who knows more about heart pacemakers: an electro cardiologist or your family doctor? As I said before, some companies tout the breadth (general-ness) of their offerings. Specialists do just the opposite. They exude laser beam focus. Can you credibly claim such focus that no competitors can? Before you answer, understand doing this can get tricky. I regularly see marketers say they specialize in something that turns out to be almost wholly generic. We have a client that manufactures and markets industrial abrasives. They don't refer to themselves as "industrial abrasive specialists." That would be uninteresting and generic, because most suppliers in this category could credibly make an identical claim. So, kudos to our client for staying away from that. Instead, they position themselves as the creators of the industry's first and only superoxalloy abrasive. (Google it.) Specialization in superoxalloys has provided them with a meaningful and relevant differentiator in their crowded segment.

Six Differentiators to avoid

By and large, stay away from these ideas; they rarely differentiate:

1. "Our People"

I just typed the phrase, "Our people make the difference," into Google. I got back 19.1 million hits, including 2.28 million videos in .037 seconds. This "differentiator" commands even less uniqueness than I could ever have imagined. And believe me, I already figured it to be quite meaningless. Millions and millions and millions talk about the amazingness of their people, hardly a differentiator. I clicked a few of the Google links, and learned that electrical contractors, funeral homes, HR consultants, trucking companies, sign language schools, non-metal door makers, bank fraud investigators and countless others tout how their people make the difference.

It is, of course, 100 percent true. No company can do anything, including make a difference without people. Are your people great? Do they do a good job, make a difference? Sure, and that's terrific. And you should absolutely reward and recognize them for that. But it's not a marketing differentiator. It just isn't. And it never will be. Go back to the list above and see if you can transform something about what your people *do* into a true differentiator.

2. "We really, really care for our customers;" "We're committed to customer success;" etc.

Today, just about every company knows to be customer-centric. It's to the point where talking about it has become meaningless, perhaps even insincere. Truth is, lots of suppliers care about their customers. And every supplier has satisfied customers who believe they receive great care and attention. This means many customers you wish to woo already believe they get great service from their current supplier. Don't insult them by implying they're getting or accepting lousy service. We once tested a message that tried to convince users of a competitive product they were "compromising" and could do better. The audience took offense and that was the last time we mentioned it.

But what if you do care more than others and you really, really, really take care of your customers? My advice: don't say it. No one will believe you. Find something else differentiating to talk about. I realize not everyone will follow my suggestion. So, if you must talk about how much you care for your customers, stick to facts. Explain what you do that's different and relevant in concrete terms. Say things like, "We ship 99.95 percent of orders in 24-hours or less . . . or, our employees (not call center subcontractors) answer our phones 24/7/365." Or pass along what others say about you (idea #6 above): "Machine shop owners rank us #1 in on-time deliveries, as reported by . . . " or: "6,274 five-star ratings and counting." Consider using direct quotes from satisfied customers by name and company.

Customer quotes and the permission to use them aren't always easy to come by. But it's usually worth going to the effort to get them. Real praise from real customers shows you care way more than simply saying those words yourself.

3. Broad product lines/service offerings

This one's simple. Those that hype a broad line, inadvertently make the point that they do not specialize. Clarification: It's okay to differentiate on being narrow and deep. But it's not okay to differentiate on broad and shallow. For instance, let's say you offer thousands upon thousands of different diamond-coated cutting tools. That shows narrowness and depth —specialization—not breadth. On the other hand, positioning yourself as broad states or implies you do everything or darned near everything well. Who would believe this?

I realize firms can be successful selling broad product lines. However, when you look closer you tend to find lots of choices in a narrow area. Take General Motors, who sells automobiles. As an aside, I once heard a marketer joke, "They'd be more successful if they became *Specific* Motors." And indeed, they have chosen to narrow their product lines in recent years. You could say GM offers a broad line: compact cars, luxury cars, pickup trucks, SUVs, minivans, crossovers, work trucks, etc. But truth is, they only sell automobiles. They don't sell jet aircraft or aquaculture feeds or hip implants. They just sell autos.

On the other hand, it can be common for distributors or resellers to offer broad lines. Take Grainger, a B2B distributor who, according to their website, is "North America's leading broad-line supplier of maintenance, repair and operating (MRO) products . . ." For the most part, they resell specialized products made by thousands of their suppliers. Grainger sells these products into a fairly broad market (MRO). At the same time, their suppliers invest money into differentiating and merchandising the utility of their products to end users. They intend to influence potential customers who stop by the Grainger website or one of their brick-and-mortar stores to choose one of their products instead of a competitor's. To be fair, Grainger positions their broad line as a differentiator using this tagline: "The one product you need, + 1.5 million more." Not bad, if you ask me. As a distributor/reseller of supplies, their "we've got whatever you need" makes sense. But for most B2B product marketers, specialization, rather than a broad offering, provides far greater opportunities to differentiate.

4. Quality

Hats off to the world of continuous improvement for improving the quality of nearly everything. Not so long ago, buyers could more easily spot differences in quality—much less so today. So, can you differentiate on quality? In many B2B markets, customers consider quality a given. In some highly technical fields, marketers may still be able to differentiate on quality, often through a proprietary process. If that describes you, give that process a relevant, differentiating name and story, and back that up with facts. Another option is to take a small part of your quality story—a technical feature of how you do what you do—and explain how this seemingly modest difference provides a direct and relevant benefit to the customer. Also understand that quality discussions and examinations often come into play in the later stages of a buying process when the prospect's people and your people are engaged one-on-one. Don't squander your early marketing efforts talking about finer points that fit better later in the discussion.

5. Adjectives

"Incredible strength!" "Unparalleled speed!" "Silky smooth." Attorneys have a word for this type of sales copy: "puffery"—a word defined as: "vague, broadly exaggerated or boastful statements about a product or service that are subjective (a matter of opinion), rather than objective (able to be measured)." I'm no attorney, but my understanding is that adjective-driven puffery is often considered legally okay, because as the lawyers argue, no reasonable person would believe it. As an example, consider this phrase: "silky smooth, stainless steel, surface." I'm guessing this phrase meets the criteria for puffery because no reasonable person would believe that steel (a metal) could ever feel like silk (a fabric). Use of puffery indicated by over-use of adjectives is almost always a bad idea.

Unlike facts, adjectives rarely differentiate. Instead of using them, I recommend you, instead, quantify the point you wish to make. For instance, don't say: "incredibly fast cutting speed." Do say: "cuts through 24mm tempered steel in less three seconds per linear foot." One other advantage of avoiding "salesy" adjectives: you'll likely sound different than your competitors. This small difference just might boost your credibility. Here's a tip on how to root out puffery. When you feel like putting an exclamation point behind a phrase, that's a good clue you've engaged in puffery. Incredible strength! Nope. Time to re-write. If a customer uses adjectives to describe your product—"lightning fast," for example—that's okay. Use their words verbatim as a testimonial. But don't you make the mistake of loading up your marketing with skepticism-inducing, adjective-laden copy of your own creation.

6. Lowest Price

I won't belabor this. And why would you want to differentiate on price anyway? Marketers sometimes do this. And in a few cases, they succeed. Very few.

CHAPTER 8

Conclusion: It pays to be different!

Have you read *Blue Ocean Strategy*? Published in 46 languages, it's one of the best-selling business books of all time. To date, it has sold more than four million copies. Read it. Or you can read this, my one sentence summary:

> To succeed in business, operate in an uncontested market.

Uncontested market? The authors make the case for differentiation. Along these lines, consider this from Ian C. MacMillan and Rita Gunther McGrath, in their *Harvard Business Review* article, "Discovering New Points of Differentiation":

> "Most profitable strategies are built on differentiation: offering customers something they value that competitors don't have."[1]

And this from an article about B2B customer brand loyalty:

> " . . . as the buyer perceives the supplier brand to be more differentiated, the perceived brand value and brand image of that brand become higher."[2]

Yes, differentiation does good things for businesses. But *how* do we do it? *How* to differentiate in B2B marketing—that's what we've discussed in *Different Rules*. Here's a one sentence summary of this book:

> To differentiate and win in B2B, play by different rules.

That's it? Yes.

Differentiation: If it's so simple, how do people get off track?

When it comes to differentiation, we've seen and made plenty of mistakes over the years. Here are a few of the most common ones:

1. Too much internal focus

Companies, especially big companies, struggle with this. Successful small companies often don't—especially ones built from scratch—firms that had to struggle to survive. Through trial and error (mostly error) they learned to put customers and the wisdom of the marketplace ahead of everything. If they hadn't, they'd be out of business. You could read an entire library shelf-full of books on how to get over the problem of too much internal focus. I have a remedy. And you're going to think it's way too naïve or simple or unsophisticated, but here goes: when in doubt, ask the customer.

2. Leaving important people out of the process

In your search for differentiation, get important internal and external people involved. And get them involved early. Who are they? Customers. Prospects. Salespeople. Technical people. Finance. Legal. Manufacturing. And many, many others. I can't give you the list. You can. Before you're too far along, sit down and make a list of all the people who can help or hinder you in your differentiation quest. Next, get their input. Get to know these people. Ask for their advice. Learn what they care about. Find out what's relevant to them. Later, when you present your differentiation plan for approval, you'll know their concerns and be able to address them. Better yet, if you did a good job understanding and engaging them early on, you may have at hand a pre-assembled army of internal and external allies ready, willing, and able to support you.

3. Thinking "best" or "better" instead of *different*

It seems simple: to differentiate, be *different*. So why do we screw this up? Fact is, we can't help it. In our society, being better or best is considered the ultimate. Being different gets frowned upon. But in marketing, because of the way the human brain works, being *different* is the ultimate. In the mind, *different* rules. I've read the neuroscience. I've studied the psychology. And yet I still struggle to remember and act unfailingly in favor of different over better or best. But I know better! And now, after reading *Different Rules*, you do, too.

4. Leaving out the *because*

Feelings and emotions matter to all people, including those who buy B2B goods. But because B2B audiences buy for a business, factors beyond feelings and emotions also get considered. And the closer you get to that buyer making the purchase, the more important and persuasive these rational reasons become. By all means, talk about product benefits. But never forget to support those with sound reasons—a *because*. It can be mindlessly easy to assume everyone understands the why behind your product and its benefits. But that's a mistake. And when it comes to rational support for what you promise, features are your friend. Use them as your *because*.

5. Complex, boring, irrelevant messaging

When you confuse or fail to connect, or when you ask people to put extra time and effort into deciphering your story, they will tune you out. And when people tune you out, you cannot and will not rule your market. Who will tune you out? Your salespeople, dealers and resellers, influencers (internal and external), the media covering your industry, and yes, customers and prospects. A proper Marketing Message overcomes this with brevity, clarity and impact. Remember, you get only a few brief seconds to grab attention and convey your difference and relevance. Do not squander those moments. And once you've engaged your audience, be ready with additional messaging that spells out the utility of your product and harnesses the power of three.

6. Poor creative execution

A great differentiation strategy doesn't guarantee success. You can't bore or confuse people into buying your product. Winning in the marketplace requires great creative execution in marketing communications. Imagine the power of creative work that brings the strategy to life, conveying your difference and relevance with the right appeal, married to interesting and compelling words and visuals. Now imagine a great strategy hitched to boring, confusing, irrelevant creative execution. It can't work. Creativity that differentiates in B2B is no game for the faint of heart. It takes knowledge, experience and commitment to excellence and funding to get your differentiation message into the market.

7. Lack of enrollment

Up until now, we've talked little of how and why to get your organization on board with a differentiation marketing strategy. And yet in B2B, the difference between getting and not getting your customer-facing team on board (enrolled), can make or break you. That's because, in B2B, many people on both the buyer and seller side become involved. This isn't so with many consumer goods. I can buy a new brand of candy bar without ever meeting or interacting with a Nestlé sales rep or technical person. But no company could ever buy something like an industrial gas turbine without interacting with multiple people in the seller's organization. Those people must be on board with what makes you different. And they must be able to convey that difference and its relevance throughout the sales process. Before you launch a differentiation campaign, take the time to enroll everyone in your organization who interacts with customers. Explain the rationale, message and content of your marketing campaign. Include internal enrollment as the final pre-launch phase of your campaign. It's an important step on the path to sending a uniquely different brand into the marketplace. Chapter Six in *The People Powered Brand* outlines a process for getting this done.

Be different. Be relevant. Be well!

I began *Different Rules* with a look at how and why being different has become so important to success in B2B marketing: too many choices, too little time.

I doubt anything has changed since you read those words. But if it has, I suspect the problem has only gotten worse. All the more reason for you to double and redouble your efforts. Be different. Be relevant. Express your difference. As Jack Trout wrote, "Every aspect of your communication should reflect your difference."

In closing, I leave you with one final thought, a question:

What makes your company, your product *different*?

Within your answers lies the seed of differentiation. Tend to it according to the rules.

Appendix
General Guidelines about Questions and Questioning

This list of 32 idea-starter questions that will work for most any interview in any market, first appeared on pages 45–50 of *The People Powered Brand.*

Start broad; then go specific. When interviewing customers, for example, begin by asking them to explain what they do in their jobs. (Later you can probe on how they feel about the company.) The pattern of general to specific should be followed the whole way through.

You're there to listen, not justify, defend or answer questions. Explain this up front, before the interview begins. Then, should someone happen to ask you a question, remind him or her of that and then turn the question around. For instance, if an employee asks you, "Is the company doing anything about issue X?" Turn it around by saying, "What's your belief about this? Are they working on issue X? What should they be doing about issue X?" See how this works?

I've found there's usually an intense opinion hidden behind a question asked of the interviewer. Don't miss out on the learning opportunity by answering the question!

Feel free to go "off script" by following the interviewee's train of thought. While your prepared questions may be very well planned, people will routinely stray from your agenda by bringing up ideas and topics that you may never have considered. (Stumbling upon something you may never have thought of is one of the main advantages of qualitative research vs. quantitative.) I tell clients, "We probably won't make it through the entire list of questions in every interview. If someone introduces an out of the box idea, we're going to put the script away and follow them." And we do. So be curious. Take advantage of the opportunity to learn things you could never have predicted by setting aside your questionnaire.

Tap into the power of "the last question" by asking at the end of *every* interview: "Is there anything else you want to tell us that we haven't asked you about?" We usually ask this question after telling the subject we are finished with the interview. Instinctively, people relax when they know they've made it through the session and feel they're no longer "on stage." And that's often when we learn about important issues we never would have considered. We are constantly amazed at what we discover with this question. In fact, we estimate that 30 to 40 percent of the insights we uncover in any project result from "the last question." Use it to your advantage by including it at the end of *every* interview.

Dig for *opinions* and *attitudes*, not *solutions* to your problem (yet). Whether the people you interview are employees, customers, prospects or industry experts, all of these people can tell you what they like and don't like. They can talk about the problems they face. They can describe how they go about performing a task. They can explain such things as: what they do in their job, what sort of tasks they perform in a typical day, what frustrates them about the way things stand, what they dream about and what happens in their job that makes it a "good day" vs. a "bad day." But, remember, they can't ordinarily explain, for instance, *why* they like beige as the color of the instrument they use every day in the laboratory. And often, they can't tell you what you should do to fix things—nor is that something you want them to tell you. That is your job!

So, avoid the temptation to ask questions such as: "If you were running the company or head of R&D, or king for a day, or if you could wave a magic wand what would you do about __________?" I've tried all of these questions and countless other variations with limited success. Eventually I figured out questions such as these, in essence, ask the respondent to do *my* job. But since the people being interviewed don't have the same skill, experience, source data and mindset as me, they can't do it. Instead, I've found people can much more easily describe their feelings and behaviors—which is precisely the information you need to know.

What to ask (sample questions). Perhaps it goes without saying, but you will need to tailor your questions, which we refer to as a "Discussion Guide," to your specific situation. Some of the questions below will work for people inside the company, some for folks outside and some for both. Here are a few to get you started as you develop your Discussion Guide:

Tell me about yourself. What do you do?

What sorts of things does your job entail? Describe a typical day.

What are the biggest trends in the industry (or product category)? What is driving those trends? What are you doing to deal with/adapt to these trends?

Where do you think this company is headed? Can you describe their vision for the future?

Tell me about the type of day when . . . everything goes perfectly (or when . . . just about everything goes wrong). What does that look like?

What is the biggest problem you face with (a key issue, competitor, etc.)?

What do you enjoy about your job? What don't you like?

If a friend (or work colleague) asked you about the company (or company X), what would you say?

What does the company (or company X) do well? What don't they do well?

Tell me a story about how you learned of a new product or company that sells/offers _________ types of products. (Insert your industry or category in the blank.) How did you come to find out about the product or company? What did you do next? (And after that?) How did things turn out?

Can you describe an instance where you tried a new product (or supplier) and things worked out really well (or didn't work out very well)? What happened?

What letter grade would you give the company (or product) when it comes to _________ ? (Quality, value for money, innovation, etc.) What caused you to give it that grade?

When it comes to _________ what company comes to mind?

What excites you/inspires you the most about the company (or this industry, or your job)?

As you think about your future, what inspires you most (or worries you most)?

Tell us a story of _________ (how you came to work for this company . . . how you first learned about _________ , etc.).

Don't be afraid to probe for deeper and more complete answers by asking follow-up questions to the respondents' comments, such as:

How important is that to you?

What is the cause of that?

What is the impact of this?

What does that mean to you?

How do you feel about that?

What would things look like if you fixed (or didn't fix) that?

You said "_________." Can you tell me more about that?

Stanley Payne's *The Art of Asking Questions* is a terrific resource for help in constructing questions. It deals mainly with quantitative surveys, but is also quite useful for qualitative research. The book is out of print, so you may need to do some digging—but the effort is well worth it.

How to conduct the interviews? Start your Brand Discovery research with the company's executive leadership, followed by other employees. Then move to people outside the company: prospects, customers, industry experts, etc.

Although I've referred to these as "one-on-one" interviews, we've found that two-on-one (two interviewers, one subject) works best. One interviewer can mainly lead the discussion; the other can mainly take notes—although both should take notes so they can later compare what they heard.

Analyzing the data for insights

Capture your initial impressions in real time. Immediately after completing an interview, create an Interview Summary to recap what you heard. Do this while the interview is still fresh in your mind. Do not try to do it later! This is hard work and you won't want to do it, but it will save you lots of time later when you must synthesize what you've heard into key findings. While you should tailor your synopsis to your specific situation, we find answering questions such as these can help you get to the heart of your findings for each individual interview:

What key ideas or themes emerged?

What did the subject care about the most?

What did you hear that surprised you?

Did any new issues come up that should be probed in later interviews?

References

PREFACE

1. Drucker, P., *The Practice of Management.* Harper & Row, 1954.

INTRODUCTION

1. Levitt, T., "Marketing success through differentiation—of anything." *Harvard Business Review,* January–February 1980.
2. Levitt, T., *Thinking About Management.* The Free Press, 1991.
3. Cialdini, R., *Influence: The Psychology of Persuasion.* HarperCollins, 2007.
4. Hallward, J., *Gimme! The Human Nature of Successful Marketing.* John Wiley & Sons, Inc., 2007.
5. Ulwick, A., "Turn customer input into innovation." *Harvard Business Review,* January 2002, 91–97.
6. Drucker, P., *Management: Tasks, Responsibilities, Practices.* Routledge, 2011.

CHAPTER 1

1. Hallward. *Gimme!*
2. Medina, J., *Brain Rules: 12 Principles for Surviving and Thriving at Work, Home and School.* Pear Press, 2014.
3. Magretta, J., "Stop competing to be the best." *Harvard Business Review*, November 30, 2011.
4. Moon, Y., *Different: Escaping the Competitive Herd.* Crown Business, 2010.
5. Bransford, J. [et al.], editors. *How People Learn: Brain, Mind, Experience, and School,* National Academy Press, 2000.
6. Magretta. "Stop competing to be the best"
7. Hallward. *Gimme!*
8. Medina. *Brain Rules*
9. Hallward. *Gimme!*

CHAPTER 2

1. Costa Hernandez, J., Wright, S., Rodrigues, F., "Attributes versus benefits: the role of construsal levels and appeal type on the persuasiveness of marketing messages." *Journal of Advertising,* 44(3) 2015, 243–253.

2. Langer, E., Blank, A., Chanowitz, B., "The mindlessness of ostensibly thoughtful action: the role of "placebic" information in interpersonal interaction." *Journal of Personality and Social Psychology*, 35(6) 1978, 553–642.

3. Carpenter, S., Glazer, R., Nakamoto, K., "Meaningful brands from meaningless differentiation: the dependence on irrelevant attributes." *Journal of Marketing Research,* 31(3) 1994, 339–350.

CHAPTER 3

1. Hallward. *Gimme!*

2. Medina. *Brain Rules*

3. Carnegie, D., *How to Win Friends and Influence People.* Cornerstone Publishing, updated ebook version, 2005.

4. Spenner, P. and Freeman, K., "To keep your customers, keep it simple." *Harvard Business Review*, May 2012, 108–114.

5. Zaltman, G., "How customers think: essential insights into the mind of the market." *Harvard Business School Press,* 2003.

6. Hauser, J. and Griffin, A., "The voice of the customer." *Marketing Science,* 12(1) Winter 1993.

CHAPTER 4

1. Trout, J. and Ries, A., "The positioning era cometh." *Advertising Age,* April 24, 1972, 35, 38.

2. Trout, J. and Ries, A., "Positioning cuts through chaos in the marketplace." *Advertising Age,* May 1, 1972, 51–53.

3. Ries, A., "Key question for brand-building: what category are we trying to create?" *Advertising Age,* January 15, 2013.

4. Bel-Art®, SP Scienceware® and Riteflow® are registered trademarks of Bel-Art–SP Scienceware, a division of SP Industries, Inc.

5. Wedgie™ is a trademark of Red Robin International.

6. SONIC® is a registered trademark of America's Drive-in Brand Properties LLC.

7. Thickburger® is a registered trademark of Hardee's Food Systems, Inc.

8. Spenner and Freeman. "To keep your customers . . ."

CHAPTER 5

1. Medina. *Brain Rules*

2. duPlessis, E., *The Advertised Mind: Ground-breaking Insights Into How Our Brains Respond to Advertising*. Millward Brown and Kogan Page Limited, 2006.

3. *Ibid.*

4. Shu, S., and Carlson, K., "When three charms but four alarms: identifying the optimal number of claims in persuasion settings." *Journal of Marketing,* 78, January 2014, 127–139.

5. Cowan, N., "The magical number 4 in short-term memory." *Behavioral and Brain Sciences*, 24, 2000, 87–185.

6. Medina. *Brain Rules*

7. Brenner, L. and Ward, A., "Accentuate the negative: The positive effects of negative acknowledgment." *Psychological Science,* 17(11) 2006, 959–962.

8. Medina. *Brain Rules*

9. Hallward. *Gimme!*

CHAPTER 6

1. Antle, J., "Message to B2B marketers: we can be creative." *Forbes* online, September 22, 2020.

2. Baack, D., Wilson, R., and van Dessel, M., "Advertising to business: does creativity matter?" *Industrial Marketing Management.* 55, May 2016, 169–177.

3. Pieters, R., Warlop, L., and Wedel, M., "Breaking through the clutter: benefits of advertisement originality and familiarity for brand attention and memory." *Management Science,* 48(6) June 2002, 765–781.

4. Stewart, D., and Koslow, S., "Executional factors and advertising effectiveness: a replication." *Journal of Advertising,* 18(3) 1989, 21–32.

5. Yang, X., and Smith, E., "Beyond attention effects: modeling the persuasive and emotional effects of advertising creativity." *Marketing Science,* 28(5) September-October 2009, 935–949.

6. Persky, J., "The ethology of *homo economicus.*" *Journal of Economic Perspectives,* 9(2) Spring 1995, 221–231.

7. Gong, Z., and Cummins, G., "Redefining rational and emotional advertising appeals as available processing resources: toward an information processing perspective." *Journal of Promotion Management,* 20(2) 2020, 277–299.

8. Johar, J. S., and Sirgy, M. J., "Value-expressive versus utilitarian advertising appeals: When and why to use which appeal." *Journal of Advertising,* 20(3) 1991, 23–33.

9. Lautman, M. R., and Percy, L. "Cognitive and affective responses in attribute-based versus end-benefit oriented advertising." *Advances in Consumer Research,* 11, 1984, 11–17.

10. Teichert T., Hardeck, D., Liu, Y. and Trivedi, R., "How to implement informational and emotional appeals in print advertisements: a framework for choosing ad appeals based on advertisers' objectives and targeted demographics." *Journal of Advertising Research,* 58(3) 2017, 363–379.

11. Akbari, M., "Different Impacts of Advertising Appeals on Attitude for High and Low Involvement Products." *Global Business Review,* 16(3) 2015, 478–493.

12. Lewis, H.G., *On the Art of Copywriting,* 3rd edition. Racom Communications, 2004.

13. Lewis, H.G., *Killer Copy in a Crisis: Words of Wisdom from Herschell Gordon Lewis.* Global Marketing Alliance, 2020.

14. Medina. *Brain Rules*

15. Copy Chasers. *Business Marketing,* June 1999.

16. Baker, S., *Systematic Approach to Advertising Creativity.* McGraw-Hill, Inc., 1979.

17. Medina. *Brain Rules*

CHAPTER 7

1. Cialdini, R., "The Consensus Principle." YouTube, https://www.youtube.com/watch?v=QppC_JIQ_0Y, accessed 12/30/2021.

2. Bohner, G., Einwiller, S., Erb, H., and Siebler, F., "When small means comfortable: relations between product attributes in two-sided advertising." *Journal of Consumer Psychology,* 13(4) 2003; 454–463.

CHAPTER 8

1. MacMillan, I., and McGrath R., "Discovering new points of differentiation." *Harvard Business Review*, July-August 1997.

2. Elmadag, B., and Peneklioglu, O., "Developing brand loyalty among SMES: is communication the key?" *Small Enterprise Research,* 25(3) 2018, 239–256.

Index

Articles

"Discovering New Points of Differentiation" (article), 117

"Innovator's Impasse" (poem), xvi, 43

"Marketing Success Through Differentiation—of Anything" (article), xi

"Message to B2B Marketers: We Can Be Creative" (article), 67

"Stop Competing to Be the Best" (article), 3

"The Magical Number 4 in Short Term Memory" (article), 55

"To Keep Your Customers, Keep It Simple" (article), 27–28

"When Three Charms but Four Alarms: Identifying the Optimal Number of Claims in Persuasion Settings" (article), 53

3-30-3-30 method, 56–61, 63, 91

A-B

A Technique for Producing Ideas (book), 59

Ad Age, 37

Advertising Age, 34

Baker, Stephen, 91

benefits, 9–15, 24, 78

Black Hole, 101

Blue Ocean Strategy (book), 117

Bohr, Niels, 29

Brain Rules (book), 2, 4, 36, 51

Brand Busters (book), 52, 80, 107

brands, purpose of, xiii

Broadbent, D.E., 55

Business Marketing, 83

C-D

Carnegie, Dale, 27–28

categories, xv, 22, 35–38, 40, 42–48, 91, 100

Cialdini, Robert, Dr., xiii, 47, 99, 106–107

Clemen, Robert, 29

combination appeals, 69–71, 74–75, 83

consumer attention, xiv, 27–28, 72

consumer research, 27

copy machine case study, 15–17, 19, 25

Cowan, Nelson Dr., 55

creativity in advertising, 66–71, 91

customer insights, 19–20, 28–31, 124–127

de Bono, Edward, Dr., 100

derived benefits, 9–15, 20, 22–24, 78

Different: Escaping the Competitive Herd (book), 2

direct benefits, 9–10, 13–15, 20, 22–24, 78

Drucker, Peter, ix, xv

du Plessis, Eric, 51

E-F-G

emotional appeals, 69–75, 111

experts, 29, 79–80, 85, 95, 124–127

facts, 4, 61, 88–90, 102–103, 113, 115–116

features, xv, 4–5, 8–12, 15–16, 20–24, 60–61, 119

Five Forces of Competition model, 2
Flesh, Rudolph, 99
Freeman, Karen, 38
Gimme! The Human Nature of Successful Marketing (book), 2, 5
Grainger, 114

H-I-J

Hallward, John, xv, 2, 4–5, 26, 65
Harvard Business Review, xv, 27, 38, 117
Henry Ford, xii
high involvement products, 71
How Customers Think (book), 29
How People Learn (book), 2
How to Win Friends and Influence People (book), 27–28
Industrial Marketing, 34
Influence: The Psychology of Persuasion (book), 47
interruption, xii–xiv
iPhone, xvi, 53, 57, 70
Jobs, Steve, xvi, 53, 57

K-L-M

Karam, David, 44–45
Krugman, H.E., 51–52
Levitt, Theodore, xi–xii
Lewis, Herschell Gordon, 76–77, 95
low involvement purchases, 71
MacMillan, Ian C., 117
Magretta, Joan, 3
marketing message template, 58–62
McGrath, Rita Gunther, 117
Medina, John Dr., 2, 4, 26, 36, 51–52, 56, 65, 83, 93, 99
message testing, 64–65, 74
Methods Improvement Team, 100, 104–105
Moon, Youngme, 2

N-O-P

new, 5–7, 48
one-sided argument, 110
persuasion, as marketing concept, xiii–xv, 53–55, 68, 99, 106–107
Porter, Michael, 2
positioning, 27, 34–35, 114
Positioning: The Battle for Your Mind (book), 35
power of because, 8, 15–19, 22, 43,103, 119–120
proof principle, 8

R-S-T

rational appeals, 69–75, 111
Red Bull, 36
relevance, 7, 25–28, 40, 50–51, 59, 66, 68, 70, 75–76, 84–86, 102, 120
Ries, Al, 34–37, 47, 66
Sbarro Pizza marketing, 44–45
self-interest, 26–29, 69
Shakespeare, 25
social media research, 27
Spenner, Patrick, 38
stock photos/images, 75, 86, 94–95
Systematic Approach to Advertising Creativity (book), 91
testing ad copy, 64–65, 74
The Advertised Mind (book), 51
The People Powered Brand (book), ix, 28, 32–33, 63, 121
Thinking About Management (book), xi–xii
Trout, Jack, 34–36, 66
Twelfth Night, 25
two-sided argument, 110

U-V-W-Y-Z

utility, xiii–xvii, 3–4, 6, 8, 13–15, 21, 31–32, 38, 40–41, 43, 45–46, 50, 52, 55, 60, 87, 102, 120
verisimilitude, 95–98, 111
Voltaire, xiii
Wheeler, John, 101
Why different?, viii–xvi, 1–7, 19, 41
Williams, Tim, xii, xiii
Winkler, Robert, 29
Young, James Webb, 59
Zaltman, Gerald, 29

About the Author

Chris Wirthwein

Different Rules is the third book on B2B brands, marketing strategy and creativity from award-winning author and marketer Chris Wirthwein. His other popular titles, which can be found in university classrooms and libraries around the world, include *Brand Busters* and *The People Powered Brand,* written with co-author Joe Bannon. Chris has offered marketing insights and commentary at universities and industry conferences, on public and commercial television and radio and on a variety of marketing and business podcasts. His viewpoints and articles have appeared in such publications as *Marketing Management, Business Marketing, CMO.com, AdAge, Adweek, Entrepreneur.com* and *Sales & Marketing Management.*

In his 30+ year career in marketing, communications, advertising and market research, Chris has worked alongside 5MetaCom colleagues and clients to develop strategies and campaigns for world-leading B2B brands in medical devices, building products, agriculture, agtech and agrifood, pharmaceuticals, additive manufacturing, industrial abrasives, aerospace, oil and gas and many other industries. His experience includes a host of well-known domestic and global B2B brands, including Owens Corning, Firestone, Honeywell, Roche Diagnostics, Elanco Animal Health, Corteva Agriscience and many others.

A golf historian, Chris worked with famed golf course designers Pete and Alice Dye over a period of 16 years to author a book about Crooked Stick Golf Club, the storied course referred to as their "original masterpiece." He is also a member of the St. Andrews Golf Club in St. Andrews Scotland. A graduate of Butler University, Chris resides in Carmel (Indianapolis), Indiana.